# Intermediate Math Puzzlers

by
Ann Fisher

illustrated by Teresa Mathis

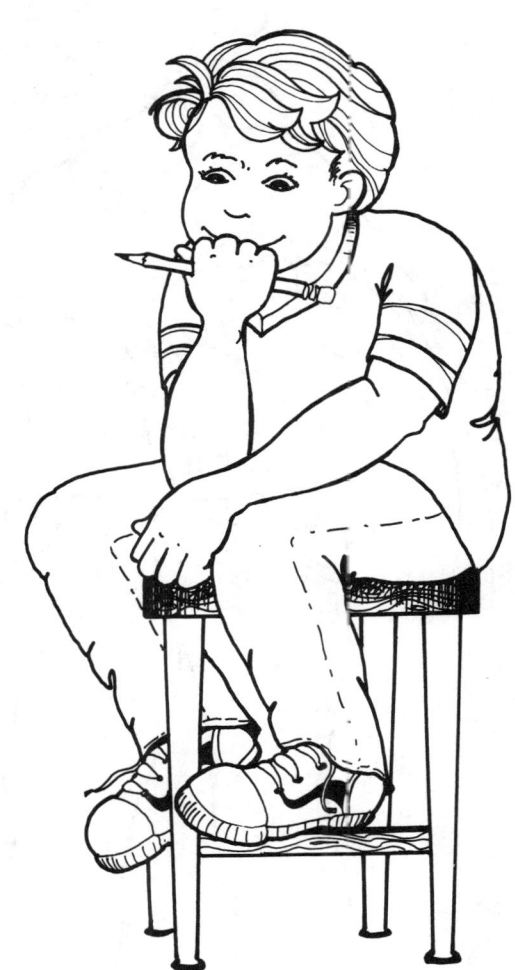

Cover by Angie Seibert

Copyright © 1994, Good Apple

ISBN No. 0-86653-813-5

Printing No. 98765

**Good Apple**
A Division of Frank Schaffer Publications, Inc.
23740 Hawthorne Boulevard
Torrance, CA 90505-5927

The purchase of this book entitles the buyer to reproduce student activity pages for classroom use only. Any other use requires written permission from Good Apple.

All rights reserved. Printed in the United States of America.

# Table of Contents

Puzzles with Addition, Subtraction, Multiplication, and Division of Whole Numbers .......1

Number Word Puzzles ................................................................................19

Puzzles with Positive and Negative Integers ................................................22

Puzzles with Prime and Composite Numbers, Factors and Multiples .............26

Exponent Puzzles ......................................................................................30

Puzzles Using Other Number Systems .........................................................33

Puzzles with Decimals, Fractions, and Percentages ......................................39

Coordinate Graphs and Map Puzzles ...........................................................57

Probability and Statistical Puzzles ................................................................63

Geometry Puzzles ......................................................................................70

Measurement Puzzles .................................................................................82

Brainteasers and Logic Puzzles ...................................................................90

Answer Key ...............................................................................................102

# To the Teacher

*Intermediate Math Puzzlers* will provide a year's worth of appealing new puzzles for your middle-grade math students. All the important math topics you teach are covered here–from whole numbers, exponents, and fractions to geometry, measurement, and logic.

Most pages in this book will fall into one of these categories:
1. Puzzles that provide fun computational practice or
2. Puzzles that emphasize problem-solving skills.

Some puzzles can provide enrichment work for independent students, others can be used as five-minute drills for the entire class, and still others can be used as group problem-solving activities.

The puzzles have been arranged into sections by broad topics, as listed in the Table of Contents. Since many puzzles cover overlapping skills, you may want to skim other parts of the book when looking for an activity to cover a specific skill. To help you, specific skills used on each page have been listed in the top right-hand corner.

Carefully preview each puzzle before presenting it to your students to be sure it is appropriate for their skill level. (Problems that are especially difficult or that extend into other areas are marked with double asterisks.) As you go over solutions, be aware that for many of the pages, only one of many solutions is provided.

Sometimes these math puzzles are more difficult to *write* than to solve. For this reason, you may want to encourage your students to write their own math puzzlers for others to solve as their abilities and interests develop. Hopefully, many of your students will discover the challenge and fun of numbers and mathematical concepts as they work the pages in this book. Happy puzzling!

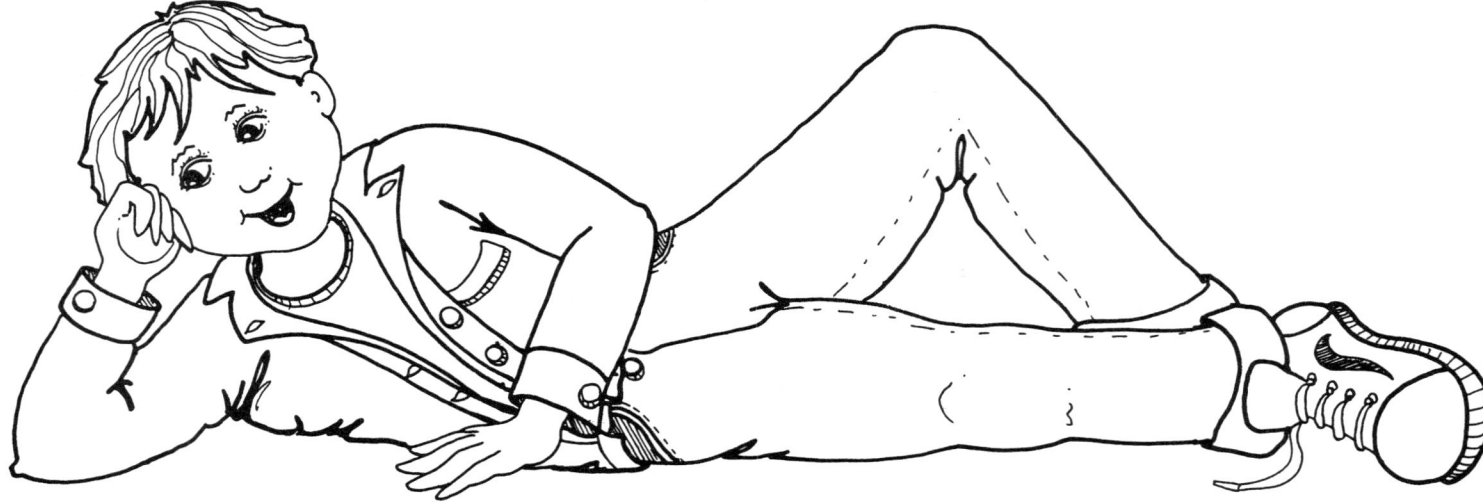

Addition and Subtraction,
Mental Computations

# Ten to Eighteen

Place the numbers 10 to 18 in the diagram so that in every row of three circles connected by lines, the sum of the end numbers minus the middle number is always the same.

Use the markers at the bottom of the page to try possible solutions. When you've found one that works, write the numbers in the circles.

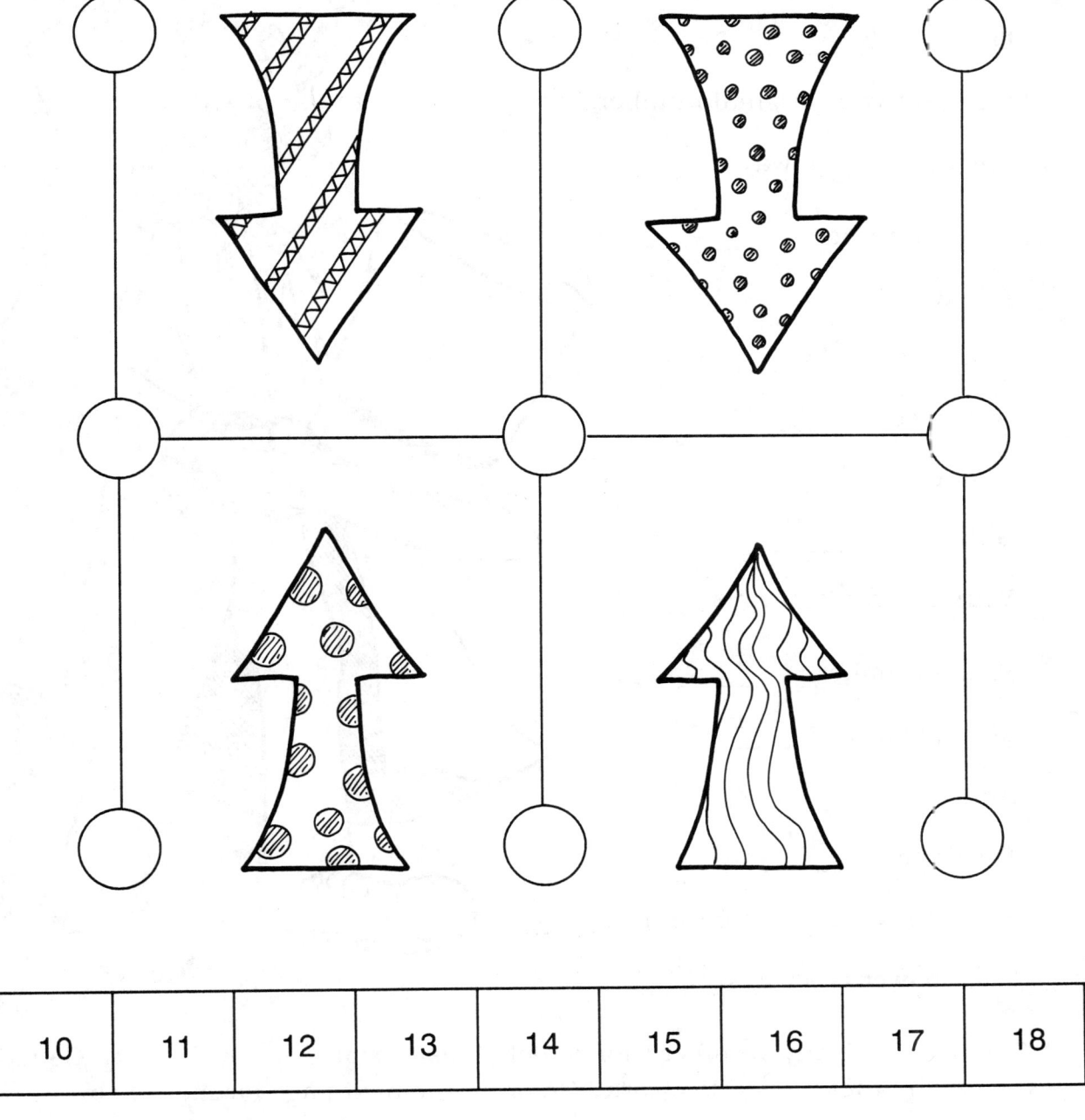

Computations with Whole Numbers,
Analyzing Formulas,
Using Algebraic Representation

# Magic Formulas

A. Pick a number. _____

   Multiply it by 3. _____

   Add 13. _____

   Subtract 7. _____

   Divide by 3. _____

   Subtract your original number. _____

   The answer is always _____ .

B. Pick a number. _____

   Add 3. _____

   Multiply by 2. _____

   Subtract 2. _____

   Divide by 2. _____

   Subtract 2. _____

   Your answer is always _____ .

C. Pick a number. _____

   Multiply it by 2. _____

   Add 18. _____

   Divide by 2. _____

   Subtract your original number. _____

   The answer is always _____ .

**Can you find any numbers for which these formulas don't work? Can you explain why these formulas work? Go on to the next page for more help.

# Magic Formulas

Let's examine the steps in Magic Formula A on the previous page and see why this formula works. Here's a step-by-step replay.

1. We pick a number. Let's call it *A*.

2. We multiply it by 3. This can be written as *3A*.

3. Next we add 13. Now we have 3A + 13.

4. Now we subtract 7. That leaves a net gain of 6. It's significant that 6 is divisible by 3, the multiplier used in step 2. This can also be written as

    3A + 13 - 7 = 3A + 6.

5. Next we divide the whole thing by 3 which will leave our original number plus 2. This can be represented in this manner:

    (3A + 6) ÷ 3 = (3A ÷ 3) + (6 ÷ 3) = A + 2

6. Finally we subtract our original number, which will always leave a difference of 2: A + 2 - A = 2

**Now try to write your own Magic Formula. Apply these same principles, using your own numbers. Then try some variations in the six steps above. Finally, analyze the other two magic formulas on the previous page. Make your own magic formulas based on those. Write your own magic formula(s) here for your teacher or classmate to try.

Addition to 60,
Mental Computations

# Sixty Sense

How many times can you score 60? Choose one number from each column below so that the sum of your three numbers is 60. Cross out each number as you use it so you don't use it again. (Do this in pencil in case you change your mind!) Finding eight or more combinations is great. Finding all twelve (and using every number) is remarkable!

| 25 | 13 | 8 |
| 28 | 17 | 10 |
| 12 | 27 | 21 |
| 19 | 18 | 15 |
| 20 | 16 | 17 |
| 36 | 11 | 18 |
| 30 | 14 | 19 |
| 22 | 30 | 20 |
| 26 | 12 | 23 |
| 17 | 15 | 24 |
| 11 | 20 | 28 |
| 23 | 24 | 31 |

Combinations you used:

_____  _____  _____

_____  _____  _____

_____  _____  _____

_____  _____  _____

Total number of 60s you found: _____

Number Play, Discovering Number Patterns,
Addition and Subtraction

# Number Pals

Part One

Here's a puzzle about number pals–palimages and palindromes. *Palimages* are pairs of numbers that have the same digits in reverse order, such as 362 and 263. *Palindromes* are numbers that read the same backwards and forwards, such as 77, 101, and 3993. If you tinker with these special numbers long enough, you will notice some interesting results.

| | |
|---|---|
| For example, start with this number: | 132 |
| Now find its *palimage* and add it: | + 231 |
| The sum is a *palindrome*: | 363 |

We'll say that the number 132 is a *one-step* number because when added to its palimage, it turns into a palindrome in just *one* step.

Here's a three-step number

$$\begin{array}{r} 263 \\ +\,362 \\ \hline 625 \\ +\,526 \\ \hline 1151 \\ +\,1511 \\ \hline 2662 \end{array}$$ palimages (1)
palimages (2)
palimages (3)
palindrome

Now look at the numbers on the next page. Can you predict which one(s) might be one-step numbers? Now check out your predictions. In the space provided, add each to its palimage. If the sum is *not* a palindrome, continue adding palimages of each sum until the result *is* a palindrome. Count the number of steps needed and write that number in the blank.

Copyright © 1994, Good Apple          5          GA1505

# Number Pals (cont'd.)

1. 431 _____ -step number

2. 462 _____ -step number

3. 251 _____ -step number

4. 541 _____ -step number

5. 564 _____ -step number

6. 901 _____ -step number

7. 384 _____ -step number

8. 287 _____ -step number

9. 749 _____ -step number

10. 603 _____ -step number

# Number Pals (cont'd.)

**Look at all the numbers that turned out to be one-step numbers.

What do they all have in common? _____

_____

Write some more one-step numbers. _____

***Work on the number 167. How many steps does it take to form a palindrome? _____

Work space:

# Number Pals (cont'd.)

Part Two

Now let's look at palimages and palindromes in subtraction. Start with a three-digit number and subtract its palimages.

```
  504
- 405
   99
```

In one step, we've found the palindrome 99. Do the same with these numbers to find other one-step numbers.

786                    372                    534

A. What do these numbers have in common? Can you make a generalization about which three-digit numbers, with their palimages, will reach the palindrome 99 in just one step? _____

Now look at these two-step numbers:

```
  882           731           863
- 288         - 137         - 368
  594           594           594
- 495         - 495         - 495
   99            99            99
```

B. Can you figure out the *rule* for numbers that turn into 99 in two steps? Experiment with more numbers until you find it. Write your *rule* here. _____

# Number Pals (cont'd.)

C. Now work through these numbers (and others, if necessary) and try to discover rules for three, four, and five-step numbers.

1. 613  2. 724  3. 926

4. 561  5. 902  6. 864

7. 371  8. 910  9. 634

D. Write your rule for three-step numbers. _____

E. Write your rule for four-step numbers. _____

F. Write your rule for five-step numbers. _____

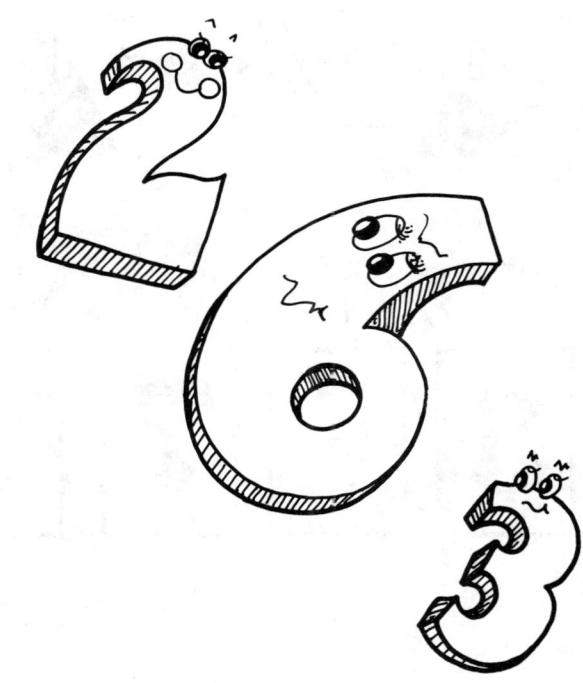

Solving Equations,
Mental Computations

# Sign In

Add signs (+, −, x, ÷, ( ) and [ ]) to make each equation true.
Examples:  24    12    6    3  =  20        (24 ÷ 12) + (6 x 3) = 20
           16     4    3    4  =  28        [(16 ÷ 4) + 3] x 4 = 28

1.  15    2    3   =  3
2.   0    7    6   =  6
3.   3    9    1    4  =  7
4.   5    9   15    3  =  9
5.   4    2    4    2  =  20
6.   4    2    4    2  =  3
7.   3    4    4    2  =  22
8.   7    6    2    4  =  30
9.  20   15    5    2  =  5
10. 20   15    5    2  =  15
**11.  1    2    3    4    5  =  6
**12. 10    9    8    7    6  =  10

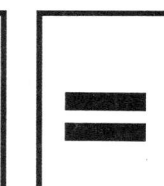

# What's the Problem?

It's time to turn the tables! Here are the *answers* to some math problems, and it's your job to write the *problems*. Here are the rules.

a. Put one digit (0 through 9) in each box.
b. Do not repeat any digit within a problem.

1.  ☐☐☐
   +  ☐☐
   ─────
    9 0 2

2.  ☐☐☐
   −  ☐☐
   ─────
      7 8

3.   ☐☐
   ×  ☐
   ─────
    7 0 2

4.  ☐☐☐
   ×   ☐
   ─────
   7 2 4 8

5.  ☐☐☐
   ×   ☐
   ─────
    4 7 7

6. ☐☐☐☐
   ×    ☐
   ─────
   2 6 , 9 1 5

Two-Digit Multipliers,
Optional Calculator Math

# Daring Dots I

We dare you to solve these problems! Cut out the dominoes under each problem. Try to arrange them into each formation shown to construct a correct multiplication problem. (A domino half with one dot represents the number 1, two dots the number 2, etc.) At your teacher's discretion, you may be allowed to check your guesses with a calculator. When you've found a solution, write it into the puzzle boxes.

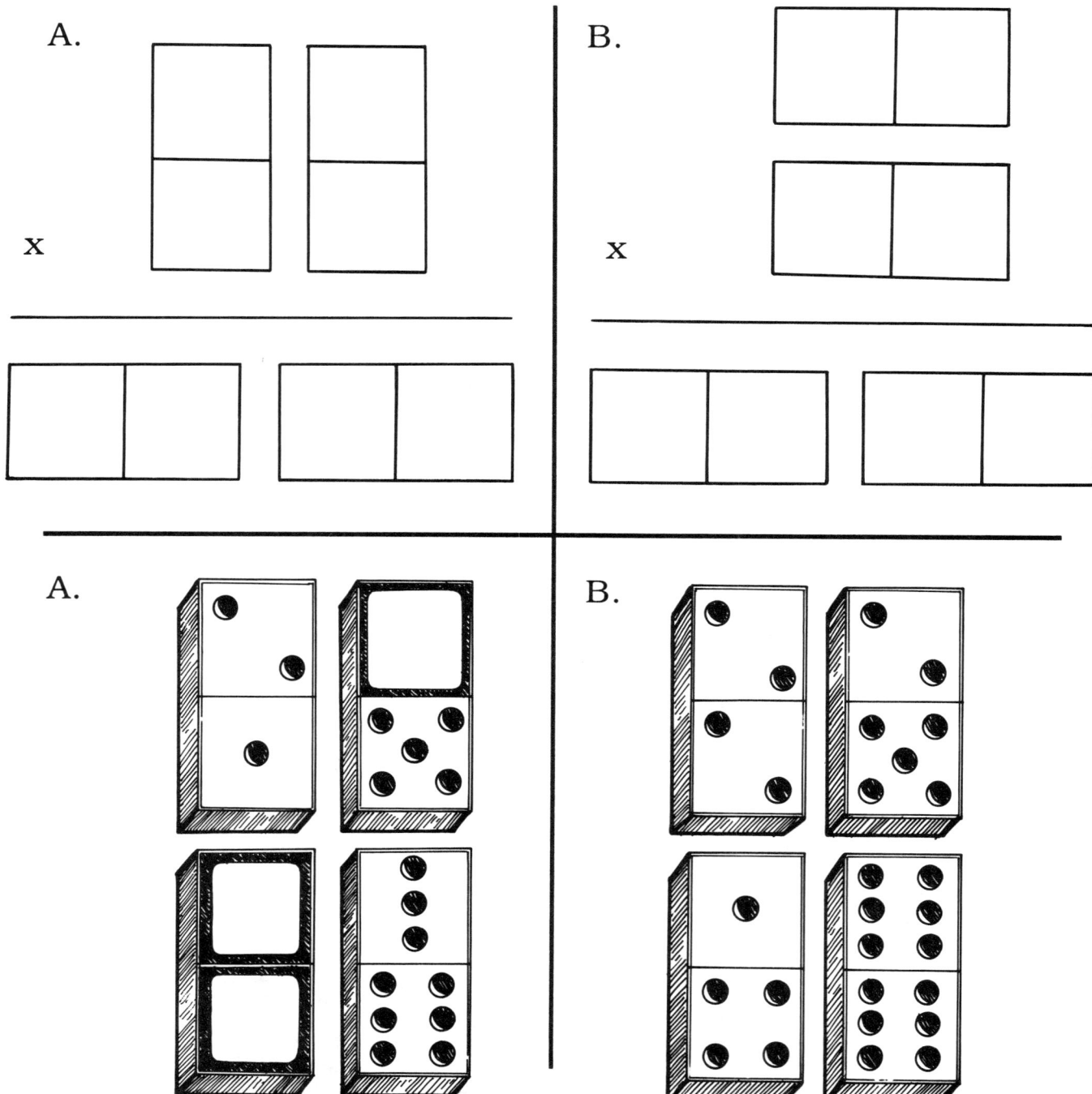

Two-Digit Multiplier and Divisor,
Optional Calculator Math

# Daring Dots II

Here are two more domino problems we dare you to solve. Again, cut out the dominoes under each problem. Try to arrange them into each formation shown to construct a correct multiplication or division problem. At your teacher's discretion, you may be allowed to check your guesses with a calculator. When you've found a solution, write it into the puzzle boxes.

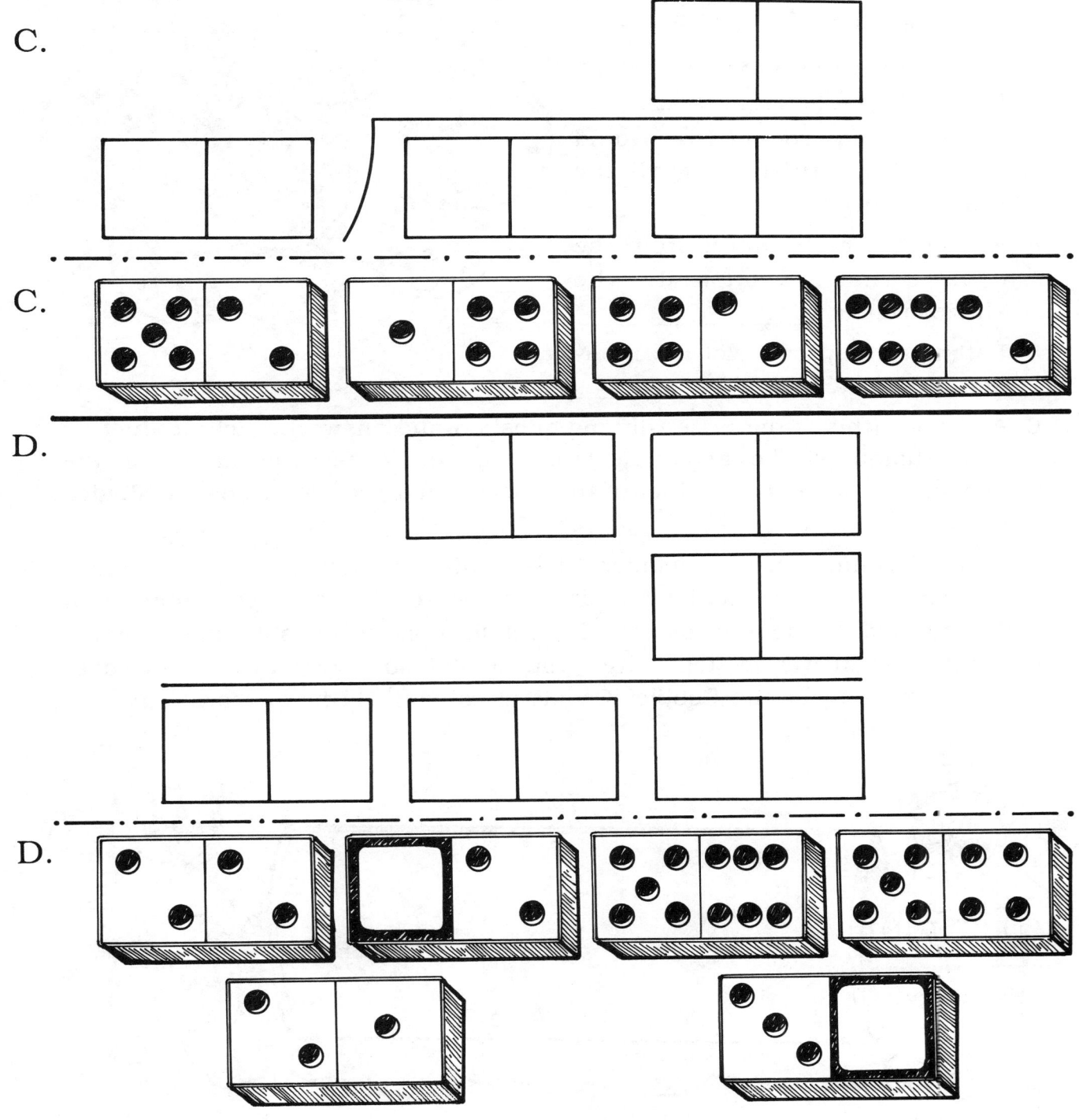

*Visualization,
Estimation, Multiplication*

# Cookie Contest

The elves at the Cobbler Cookie Company have offered a free year's supply of their Chunky, Crunchy Chocolate Chip cookies to anyone who can get this problem right. Here's what you do:

1. Figure out which of the numbered cookies are partially *under* the shaded cookie.

2. Write the numbers from those cookies here. _____

3. Estimate the product that would result from multiplying all the numbers in step 2. _____

4. Now write the numbers from the cookies that are *not* under the shaded cookie. _____

5. Estimate the product of the numbers in step 4. _____

6. And now, (this is the prize-winning question and answer) which product do you estimate will be larger, the product of the cookies that are under the shaded cookie or the product of the cookies that are not under the shaded cookie? _____

7. Now determine if you're a winner. Go back and carefully draw completed circles for each cookie. Go back and see if you selected the right numbers in steps 2 and 4. Then do the actual multiplication for steps 3 and 5. Finally, check your answer on step 6. Were you right? If so, to claim your prize cookies, watch out for the Cobbler Cookie elves who will be appearing in your neighborhood soon!

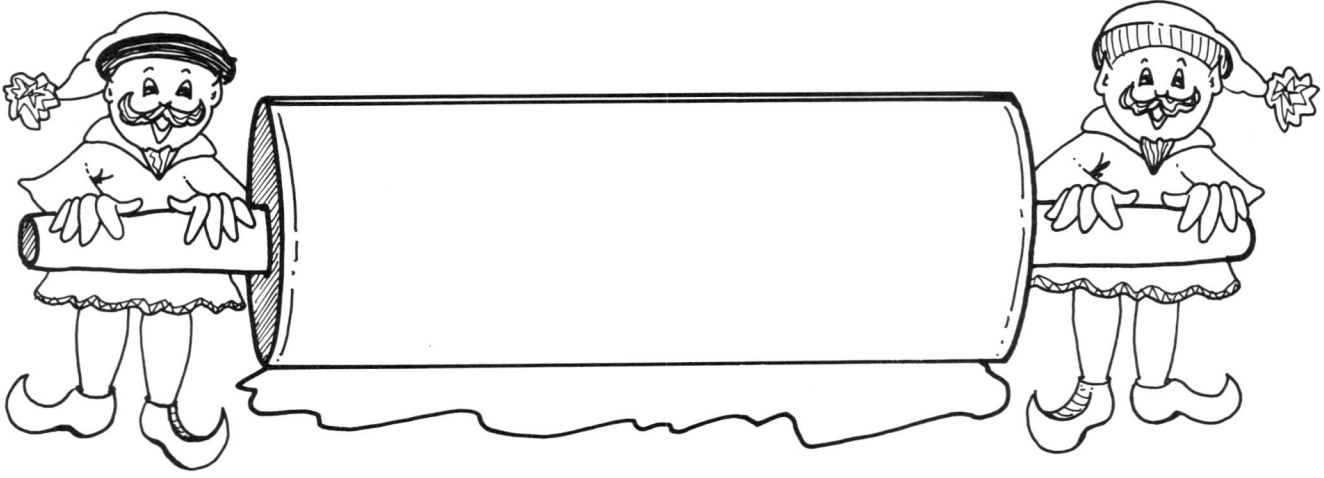

Division, Logic,
Prime Numbers,
Mental Computations

# Number Clues

Play math detective to find each number described below. Use the work space to list possible answers, and then cross out those you eliminate. Write your final answer in each blank.

| Clues | Work Space |
|---|---|
| 1.<br>　a. This number when divided by 4 leaves a remainder of 3.<br>　b. This is a two-digit number less than 50.<br>　c. The sum of its two digits is 10.  | <br><br><br>Answer: _____ |
| 2.<br>　a. This number is divisible by 7.<br>　b. It is an odd number between 50 and 100.<br>　c. The sum of its digits is also an odd number. | <br><br><br>Answer: _____ |
| 3.<br>　a. This number is *not* divisible by 2, 3, 4, or 7.<br>　b. This number is between 50 and 60.<br>　c. It is *not* prime. | <br><br><br>Answer: _____ |
| 4.<br>　a. This number when divided by 7 leaves a remainder of 5.<br>　b. It is a prime number less than 100.<br>　c. The sum of its numbers is 7. | <br><br><br>Answer: _____ |
| 5.<br>　a. This is the largest number under 100 that can be written as the product of two primes. | <br><br><br>Answer: _____ |

# Disappearing Digits

One- and Two-Digit Divisors

These division problems were finished, but now some of the numbers have disappeared. Can you figure them out? Write the correct number in each empty box.

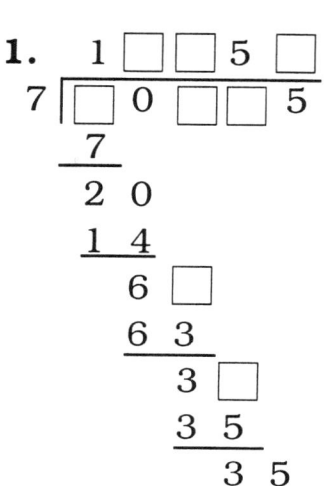

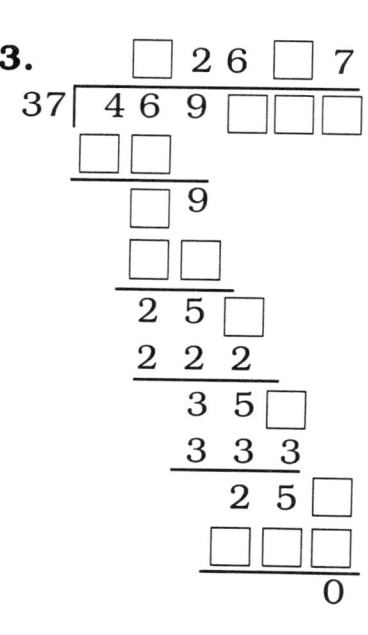

Logic, +, -, x, and ÷

# Logical Letters

Each letter in the diagram represents a different number from 1 to 9. Use the clues to figure out which letter represents each number. Write your answers in the diagram.

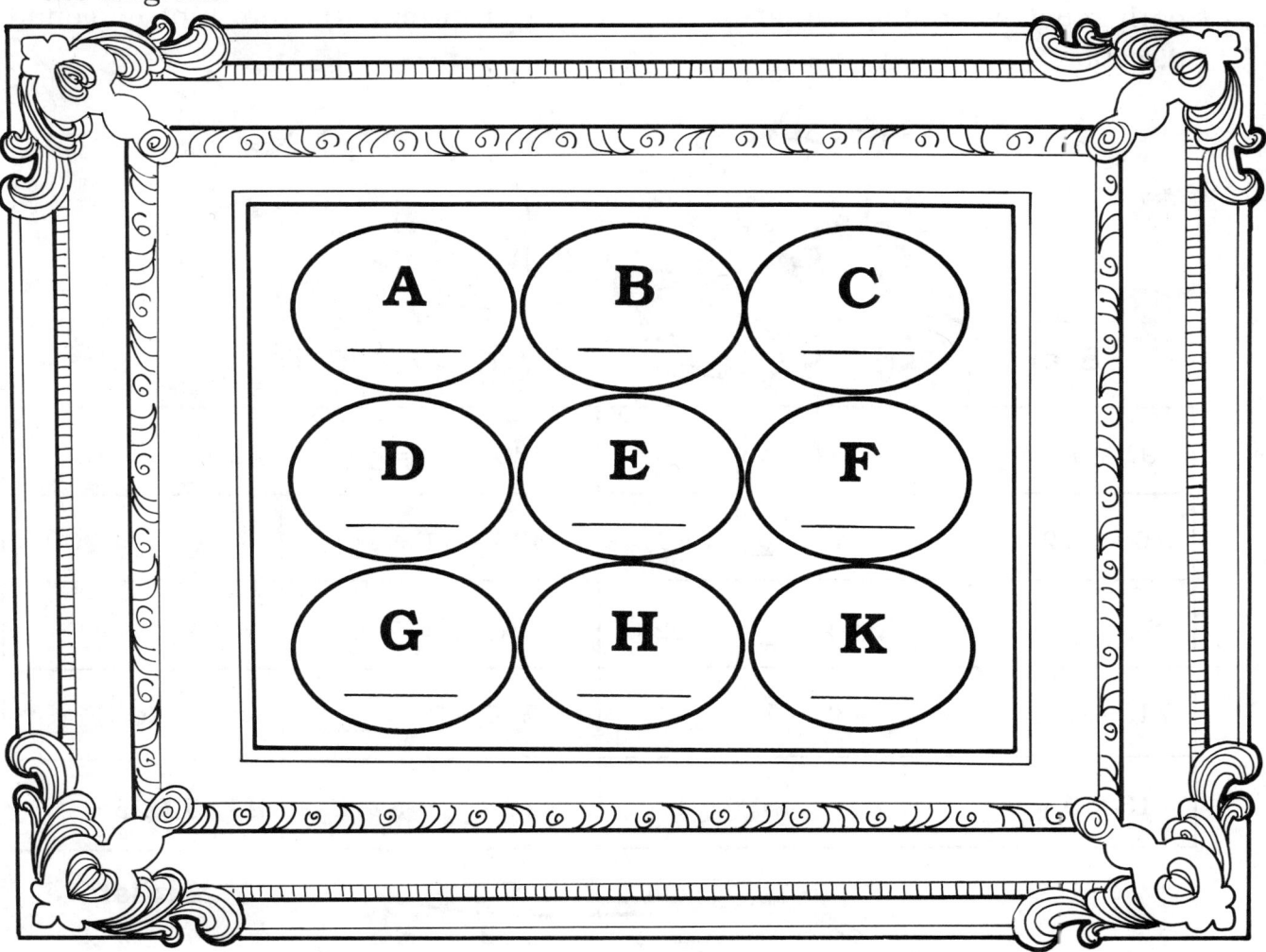

1. There are no even numbers in the first column.
2. The smallest number appears in the second row and also in the first column.
3. A + B = C
4. B and C are both divisible by A.
5. B + D = G
6. (G + D + B) - C = E
7. The sum of the numbers in the second column is 19.
8. F is larger than K.

Two-Digit Multipliers and Divisors, Patterns

# A"Maze"ing Mathematics I

First find the answer to each problem and write it in the box. Then study your answers to find some rule or pattern that will connect Start and Finish. The path must move one box at a time in a vertical or horizontal (not diagonal) direction.

**Start**

| 91 ÷ 13 = ___ | 504 ÷ 36 = ___ | 420 ÷ 20 = ___ | 6 x ___ = 150 |
| --- | --- | --- | --- |
| 108 ÷ 12 = ___ | 24 x ___ = 840 | 308 ÷ 11 = ___ | 30 x ___ = 900 |
| 8 x ___ = 416 | 31 x ___ = 1302 | 58 x ___ = 2320 | 612 ÷ 17 = ___ |
| 715 ÷ 13 = ___ | 882 ÷ 18 = ___ | 5520 ÷ 92 = ___ | 960 ÷ 20 = ___ |
| 13 x 5 = ___ | 504 ÷ 9 = ___ | 52 x ___ = 3276 | 1750 ÷ 25 = ___ |

**Finish**

Number Words

# False Advertising

Carefully read through this "fake ad" to find words for the numbers from 1 to 10. Hidden words may appear within words or be parts of two or three words.

Examples: We quo**te n**ew prices daily. You've d**one** well.

### It's the Amazing, Collapsible Umbrella!

How often have you wanted to use an umbrella but didn't because it was too bulky and inconvenient? What you need is now available! You've seen them in use at Niagara Falls, in rain forests, and yes, even in hurricanes. And now what you need is being produced in the U.S.A.! It's the Amazing Collapsible Umbrella produced by the Sta-Dri Company in Essix, New Mexico. This umbrella has been tested in the rain in every state in the nation. Because of new technology using special pulleys with reels of plastic fibers, each of our umbrellas stretches to a height of 4 feet and shrinks to a size that will slip into your back pocket. It won't be long before you'll see our umbrellas on every block. Get in on the latest fad. Don't delay–order yours today! (Endorsed by world champion surfer, Whet Waver, who says, "If I've ever found a product for keeping dry, this is it!")

\*\*There are at least three more mathematical terms hidden in the ad above. Can you find and circle them?

Number Words, Addition

# Spelling Test

What? A spelling test in math class? That's right! Here's what you need to do.

- If the number word is spelled incorrectly, cross it out and write it correctly in the blank.
- If the number word is spelled correctly, write the number for it in the scoring column.
- Then total your score and write it in the blank.

|  | **Corrections** | **Score** |
|---|---|---|
| A. thirty | _____ | _____ |
| B. ninty | _____ | _____ |
| C. thousend | _____ | _____ |
| D. eighteen | _____ | _____ |
| E. twelfe | _____ | _____ |
| F. thirteen | _____ | _____ |
| G. elevin | _____ | _____ |
| H. four | _____ | _____ |
| I. twenty | _____ | _____ |
| J. hunderd | _____ | _____ |
| K. forteen | _____ | _____ |
| L. fifteen | _____ | _____ |

Total Score: _____

Number Words, Check Features

# Check It Out

List at least twelve errors on this check.

---

**2034**

Joe Schmoe
1234 Fivth St.
Hartford Michigen

February 30, 19 94

Pay To The Order Of  _The Electric Companie_   $ 148.25

_One hunderd fowrty eihgt dollers and 35/100_

National Bank
Kalamazoo, MJ

Signature  _Jerry Schmoe_

---

1. _____
2. _____
3. _____
4. _____
5. _____
6. _____
7. _____
8. _____
9. _____
10. _____
11. _____
12. _____

Adding Positive,
Negative Integers

# Box Baffler

Place the numbered markers at the bottom of the page into the empty boxes so that the sum of each row or column matches the number shown outside the large box. Do not repeat any number within the same row or column. When you have a solution, write the numbers from the markers into the puzzle boxes.

|   |   |   |   |   | |
|---|---|---|---|---|---|
|   |   | 4 |   | -4 | **-10** |
| 2 | -7 |   | 5 |   | **-5** |
|   |   | 1 |   |   | **0** |
|   | 5 | 2 |   |   | **5** |
| -3 |   | -2 |   | 6 | **10** |
| **-10** | **-5** | **0** | **5** | **10** | |

| -8 | -7 | -6 | -5 | -5 | -5 | -1 |
|---|---|---|---|---|---|---|
| 0 | 1 | 3 | 3 | 4 | 8 | 9 |

Addition and Subtraction,
With Positive and Negative Integers

# More or Less

Move through this maze as quickly as you can. Here are the rules.

- You may move one square at a time to any square that is connected vertically or horizontally. Diagonal moves are not allowed.
- The square you move into must contain a number that is 4 less or 3 more than the square you are on.

**Start**

| 1 | -3 | -6 | -2 | -1 | 1 | 3 |
|---|----|----|----|----|----|----|
| 3 | 0 | 4 | -3 | 0 | 4 | -2 |
| 5 | -4 | -8 | -6 | -4 | -1 | 2 |
| -1 | 0 | -5 | -2 | -1 | -4 | 5 |
| 1 | -3 | -6 | -9 | -5 | -7 | 1 |
| -3 | -7 | -2 | 0 | 2 | -3 | 4 |
| -13 | -11 | -8 | -5 | -2 | 1 | 7 |
| -10 | -7 | -4 | 0 | 1 | 6 | 10 |

**Finish**

Finding Possible Solutions by
Adding Positive and Negative Integers

# Boswell's Beanbags

At Boswell's Carnival Booth, you can buy five beanbags for a buck. If you get all five bags through the holes in this board, you can win these prizes.

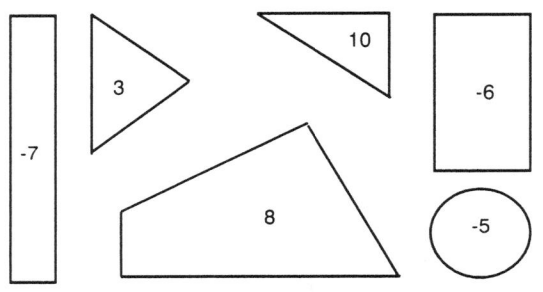

| -36 = T-shirt | 0 = elephant ear |
| -32 = whistle | 4 = softball |
| -21 = popcorn | 10 = pen |
| -9 = jump rope | 15 = baseball cards |
| -7 = visor | 30 = baseball cap |
| -1 = stuffed llama | 32 = teddy bear |

As you look carefully at the prize list, you begin to wonder if all the scores are possible.
• On the sign above, draw a line through the scores that are *not* possible.
• On the lines below, list the scores that *are* possible along with the combinations of holes needed to reach each score.

Remember: All five beanbags must go through holes in the board. More than one bag could go through any hole.

Multiplication and Division with Positive and Negative Integers

# Bingo Banger

Can you score a bingo? Find five squares in a row (vertically, horizontally, or diagonally) that have the same answer. Write your answer to each problem inside the box; then draw a line through the five squares that form the bingo.

| B | I | N | G | O |
|---|---|---|---|---|
| 2425 ÷ -25 | -388 ÷ 4 | 23 x -4 | 2944 ÷ -32 | -6 x -4 x -4 |
| -1666 ÷ 17 | -6 x 4 x 4 | -14 x 7 | 2304 ÷ -24 | 12 x -8 |
| 31 x -3 | -776 ÷ 8 | (288 ÷ 12) x -4 | 2 x 3 x -17 | -5 x 4 x 5 |
| 15 x -6 | -2 x 3 x 16 | 392 ÷ -4 | -4 x 12 x -2 | -2 x 47 |
| -1728 ÷ 18 | 7776 ÷ -81 | 19 x -5 | 612 ÷ -6 | 2 x 7 x -7 |

Factors, Multiples, Primes,
Following Directions

# Riddle Rattler

Why did Kyle take his math book to the doctor? Follow the directions below to find out.

| 0 | 1 | 2 | 3 | 4 | 5 | 6 | 7 | 8 | 9 | 10 | 11 | 12 | 13 | 14 | 15 | 16 | 17 | 18 | 19 | 20 |
|---|---|---|---|---|---|---|---|---|---|----|----|----|----|----|----|----|----|----|----|----|
| S | C | E | G | I | L | N | O | R | M | E  | U  | J  | S  | Y  | L  | M  | E  | B  | W  | O  |

| 21 | 22 | 23 | 24 | 25 | 26 | 27 | 28 | 29 | 30 | 31 | 32 | 33 | 34 | 35 | 36 | 37 | 38 | 39 | 40 |
|----|----|----|----|----|----|----|----|----|----|----|----|----|----|----|----|----|----|----|----|
| T  | U  | F  | A  | R  | P  | Y  | H  | O  | N  | T  | K  | V  | A  | P  | M  | L  | O  | S  | K  |

1. Read all the directions before doing anything.
2. When you cross out a number, also cross out the letter in the box below it.
3. Cross out all multiples of 10.
4. Cross out all factors of 24.
5. Cross out all multiples of 11.
6. Cross out all multiples of 8.
7. Cross out all factors of 30.
8. Cross out all prime numbers that remain.
9. Cross out all multiples of 7 that remain.
10. Reverse the remaining letters and write them in the blanks below.
11. Don't do numbers 3 and 7 of the directions.

Answer:

Because it had __ __  __ __ __ __  __ __ __ __ __ __ __ __

# Prime Path

Prime Numbers, Addition

A. Find a path through this maze where both numbers on each side of the line are prime numbers. The line has been started for you.

**Start**

| 2 | 46 | 8 | 31 | 41 | 9 | 10 | 24 |
|---|----|----|----|----|----|----|----|
| 3 | 5 | 11 | 13 | 17 | 13 | 6 | 27 |
| 22 | 29 | 7 | 14 | 15 | 5 | 29 | 4 |
| 16 | 51 | 25 | 28 | 27 | 37 | 19 | 21 |
| 13 | 26 | 53 | 22 | 47 | 53 | 12 | 20 |
| 34 | 59 | 52 | 18 | 37 | 71 | 23 | 57 |
| 11 | 47 | 67 | 43 | 32 | 68 | 61 | 7 |
| 49 | 33 | 61 | 63 | 57 | 39 | 53 | 59 |

**Finish**

B. Find the 2 x 2 arrangement of numbers in the maze with the highest total. Outline it and write the total here. _____

C. Find the 2 x 2 arrangement of numbers with the lowest total. Outline it and write the total here. _____

Prime Numbers, Addition

# Prime Time

Using the list of prime numbers given, try to write each even number below as the sum of two primes. See if it is always possible, or sometimes impossible. Examples: 8 = 3 + 5
26 = 13 + 13

Prime numbers from 1 to 100:
2   3   5   7   11   13
17  19  23  29  31   37
41  43  47  53  59   61
67  71  73  79  83   89
97

1. 20 = _____
2. 30 = _____
3. 40 = _____
4. 6 = _____
5. 36 = _____
6. 48 = _____
7. 54 = _____
8. 66 = _____
9. 78 = _____
10. 84 = _____
11. 90 = _____
12. 100 = _____
13. 124 = _____
14. 134 = _____
15. 148 = _____

** Try to write other even numbers as the sum of two primes. What conclusion can you draw? _____

Logic, Multiples, Addition

# Number Fits

Fit each number at the bottom of the page into one blank below. Move the markers around to try various solutions. Then write your final solution in the blanks.

Three multiples of 6   __18__  __36__  __42__

Three multiples of 7   __21__  __28__  __49__

Three odd numbers   __13__  __27__  __29__

Three numbers whose sum is 40   __6__  __14__  __20__

| 6 | 13 | 14 | 18 | 20 | 21 | 27 | 28 | 29 | 36 | 42 | 49 |

Scientific Notation, Place Value

# Exponent Express

Can you get from the start to finish in this maze? You may move from the first box to any box that touches it (diagonally, horizontally, or vertically) *if* the answer to the problem is *greater* than one million (1,000,000).

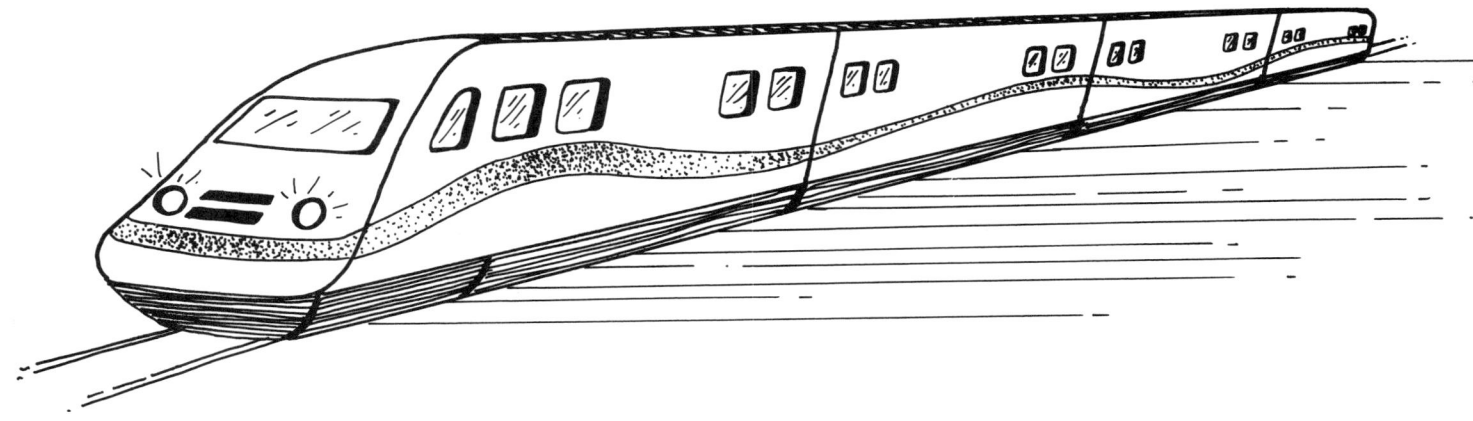

**Start**

| | | | |
|---|---|---|---|
| $10^6 + 1$ | $10^7 - 10^6$ | $10^2 \times 10^2$ | $10^5$ |
| $10^4 + 10^5$ | $10^2 \times 10^2 \times 10^1$ | $10^3 \times 10^4$ | $10^3 + 10^3$ |
| $10^5 + 10^5$ | $10^6 - 10^0$ | $10^5 + 10^2$ | $10^7$ |
| $10^4$ | $10^3 \times 10^3 \times 10^1$ | $10^6 + 10^1$ | $10^3 \times 10^2$ |
| $10^7 - 10^4$ | $10^6$ | $10^7 - 10^7$ | $10^6 \times 10^0$ |
| $10^4 \times 10^1$ | $10^7 \times 10^0$ | $10^8 - 10^7$ | $10^4 + 10^4$ |
| $10^3 \times 10^2 \times 10^0$ | $10^4 + 10^4$ | $10^6 - 10^1$ | $10^3 \times 10^2 \times 10^2$ |

**Finish**

Logic, Computations with Exponents
Optional Calculator Math

# Exponents for Experts I

Put each of the numbers from 1 through 9 into the exponent boxes to make each equation true. Use each number only once. If you like, cut out the markers at the bottom of the page and move them from place to place as you try various solutions. Some exponents appear to go in more than one place, but you must find a solution where every exponent from 1 through 9 is used and every equation is true. Write your final answers in the boxes. (Your teacher may allow you to use a calculator to help solve this puzzle.)

1. $5^{\square} + 8^{\square} = 633$

2. $6^{\square} - 2^{\square} = 88$

3. $2^{\square} + 3^{\square} = 499$

4. $1^{\square} \times 9^{\square} = 3^{\square}$

| 1 | 2 | 3 | 4 | 5 | 6 | 7 | 8 | 9 |

Logic, Computations with Exponents,
Optional Calculator Math

# Exponents for Experts II

Put each of the numbers from 1 through 9 into the exponent boxes to make each equation true. Use each number only once. If you like, cut out the markers at the bottom of the page and move them from place to place as you try various solutions. Some exponents appear to go in more than one place, but you must find a solution where every exponent from 1 through 9 is used and every equation is true. Write your final answers in the boxes. (Your teacher may allow you to use a calculator to help solve this puzzle.)

1. $9^{\square} < 8^2$

2. $2^7 < 3^{\square} < 4^4$

3. $7^3 < 2^{\square}$

4. $5^6 > 6^{\square} > 7^3$

5. $5^{\square} < 2^8$

6. $6^{\square} < 4^3$

7. $3^{\square} > 4^3$

8. $5^3 < 2^{\square} < 2^{\square}$

| 1 | 2 | 3 | 4 | 5 | 6 | 7 | 8 | 9 |

Analyzing a Number System

# Mystery Math

Study these addition "facts" for a moment.

    1 + 1 = 2    6 + 6 = 12    12 + 2 = 2
    2 + 6 = 8    7 + 7 = 2    12 + 12 = 12

Believe it or not, these unusual facts are based on a well-known object. Can you name it? _____

Once you know the object, you should easily be able to complete this addition table.

| +  | 1 | 2 | 3 | 4 | 5 | 6  | 7 | 8 | 9 | 10 | 11 | 12 |
|----|---|---|---|---|---|----|---|---|---|----|----|----|
| 1  | 2 |   |   |   |   |    |   |   |   |    |    |    |
| 2  |   |   |   |   |   | 8  |   |   |   |    |    |    |
| 3  |   |   |   |   |   |    |   |   |   |    |    |    |
| 4  |   |   |   |   |   |    |   |   |   |    |    |    |
| 5  |   |   |   |   |   |    |   |   |   |    |    |    |
| 6  |   |   |   |   |   | 12 |   |   |   |    |    |    |
| 7  |   |   |   |   |   |    | 2 |   |   |    |    |    |
| 8  |   |   |   |   |   |    |   |   |   |    |    |    |
| 9  |   |   |   |   |   |    |   |   |   |    |    |    |
| 10 |   |   |   |   |   |    |   |   |   |    |    |    |
| 11 |   |   |   |   |   |    |   |   |   |    |    |    |
| 12 |   | 2 |   |   |   |    |   |   |   |    |    |    |

**Which number in the chart above acts like zero in base ten? _____

Writing Base Two Numerals,
Using a Code

# Base Two Clues

Remember:

In base two, or the binary system, only the digits 0 and 1 are used. A portion of a simple place-value chart would look like this:

| $2^4$ | $2^3$ | $2^2$ | $2^1$ | $2^0$ |
|---|---|---|---|---|
| (16) | (8) | (4) | (2) | (1) |
|  | 1 | 0 | 1 | 1 |

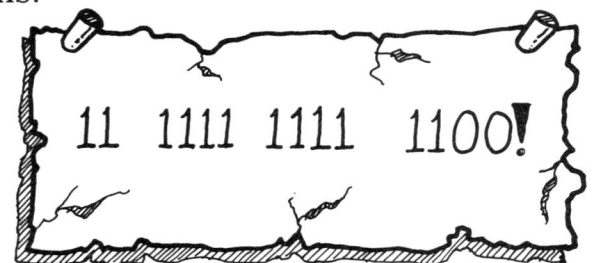

The number in the chart, $1001^2$, would equal 11 in base ten (1 one + 1 two + 1 eight.)

Finish this secret code for writing messages using base two. Imagine that A = 1, B = 2, . . . Z = 26 in base ten. Convert those numbers to base two numerals. A, B, and Z have been done to help you.

|   | Base Ten | Base Two |   | Base Ten | Base Two |
|---|---|---|---|---|---|
| A | 1 | 1 | N | 14 |  |
| B | 2 | 10 | O |  |  |
| C |  |  | P |  |  |
| D |  |  | Q |  |  |
| E |  |  | R |  |  |
| F |  |  | S |  |  |
| G |  |  | T |  |  |
| H |  |  | U |  |  |
| I |  |  | V |  |  |
| J |  |  | W |  |  |
| K |  |  | X |  |  |
| L |  |  | Y |  |  |
| M |  |  | Z | 26 | 11010 |

**Now write a short message to a classmate using your base two code. See if he or she can read your message. Have him or her send a reply, also in base two.

# A Weighty Situation

Using Base Three, Addition, Subtraction

Trenton Trigood sells tripods, tricycles, and a trio of tropical fruits. When he sells his fruits, he uses an old-fashioned balance scale to weigh them. Trenton uses only four weights–1 pound, 3 pounds, 9 pounds, and 27 pounds.

Trenton claims he can weigh any quantity of fruit to the nearest whole pound up to 40 pounds with just these four weights. Here are two examples of how Trenton weighs his fruit.

1. If Trenton wants to weigh 2 pounds of mangos, he puts the 3-pound weight on one side of the scale; he puts the 1-pound weight on the other side and adds fruit until it balances. (3 - 1 = 2)
2. To weigh 12 pounds of grapefruit, he puts the 9-pound and 3-pound weights on one end of the scale and adds fruit to the other side until it balances. (9 + 3) = 12

Fill in the rest of this chart to show how Trenton can weigh other amounts of fruit.

| Pounds to Be Weighed | Weights to Use | Pounds to Be Weighed | Weights to Use |
|---|---|---|---|
| 1 | 1 | 21 | 27 - 9 + 3 |
| 2 | 3 - 1 | 22 | 27 - 9 + 3 + 1 |
| 3 | 3 | 23 | 27 - 3 - 1 |
| 4 | 3 + 1 | 24 | 27 - 3 |
| 5 | 9 - 3 - 1 | 25 | 27 - 3 + 1 |
| 6 | 9 - 3 | 26 | 27 - 1 |
| 7 | 9 - 3 + 1 | 27 | 27 |
| 8 | 9 - 1 | 28 | 27 + 1 |
| 9 | 9 | 29 | 27 + 3 - 1 |
| 10 | 9 + 1 | 30 | 27 + 3 |
| 11 | 9 + 3 - 1 | 31 | 27 + 3 + 1 |
| 12 | 9 + 3 | 32 | 27 + 9 - 3 - 1 |
| 13 | 9 + 3 + 1 | 33 | 27 + 9 - 3 |
| 14 | 27 - 9 - 3 - 1 | 34 | 27 + 9 - 3 + 1 |
| 15 | 27 - 9 - 3 | 35 | 27 + 9 - 1 |
| 16 | 27 - 9 - 3 + 1 | 36 | 27 + 9 |
| 17 | 27 - 9 - 1 | 37 | 27 + 9 + 1 |
| 18 | 27 - 9 | 38 | 27 + 9 + 3 - 1 |
| 19 | 27 - 9 + 1 | 39 | 27 + 9 + 3 |
| 20 | 27 - 9 + 3 - 1 | 40 | 27 + 9 + 3 + 1 |

**If Trenton were to follow the same pattern and add the next larger weight to his set, how much would it weigh? 81

# A Basic Baffler

**Comparing Numbers in Bases 2, 3, 5, and 10**

This is your basic baffler–a puzzle using base 2, base 3, base 5, and base 10. Your job is to find the number in each line of the chart that is not equivalent to the other three numbers. When you find it, circle it and write the letter above it in the blank below. Then unscramble the letters to spell a two-word message.

|   | Base 2 | Base 3 | Base 5 | Base 10 |
|---|---|---|---|---|
| 1. | B 1000 | L 20 | U 13 | E 8 |
| 2. | G 10001 | O 122 | N 32 | E 18 |
| 3. | K 101 | N 20 | O 10 | T 5 |
| 4. | C 1011 | U 102 | E 22 | D 11 |
| 5. | D 11000 | R 202 | A 40 | W 20 |
| 6. | G 1101 | O 111 | L 24 | D 13 |
| 7. | W 1010 | I 100 | L 14 | D 9 |
| 8. | Z 10000 | O 120 | N 31 | E 16 |

Letters: __ __ __ __ __ __ __ __

Unscrambled letters: __ __ __ __ __ __ __ __

Roman Numerals I Through M

# There's No Place Like Rome

Here's a puzzle to test your skill with Roman numerals. Convert each clue below either to or from a Roman numeral and fill in the blanks in the crossword with the new number. Two clues have been done for you.

**Across**
1. MMMDCLII
4. MCMXLI
8. CCIX
10. DXXXII
11. LXXXIII
12. 1200
14. LX
15. 2210
17. 1964
19. 607
21. LXXI
23. 7
24. X
25. CCCLXXXIX
27. CMXLVI
28. MMDCCLI
29. MXXVIII

**Down**
1. MMMCCLXXXI
2. DCIII
3. LIX
5. XCV
6. CDXXXVI
7. MCCVII
9. 757
12. 3105
13. 212
15. 1400
16. 12
18. MMDCCXXXII
20. MMLXVIII
22. CLXXXVII
24. CXLII
26. XCV
27. XC

Remember
M = 1000
D = 500
C = 100
L = 50
X = 10
V = 5
I = 1

Roman Numerals Through C,
Multiplication

# Roman Multiplication

Write the answer to each Roman numeral problem in Roman numerals.

**Remember**

I = 1
V = 5
X = 10
L = 50
C = 100

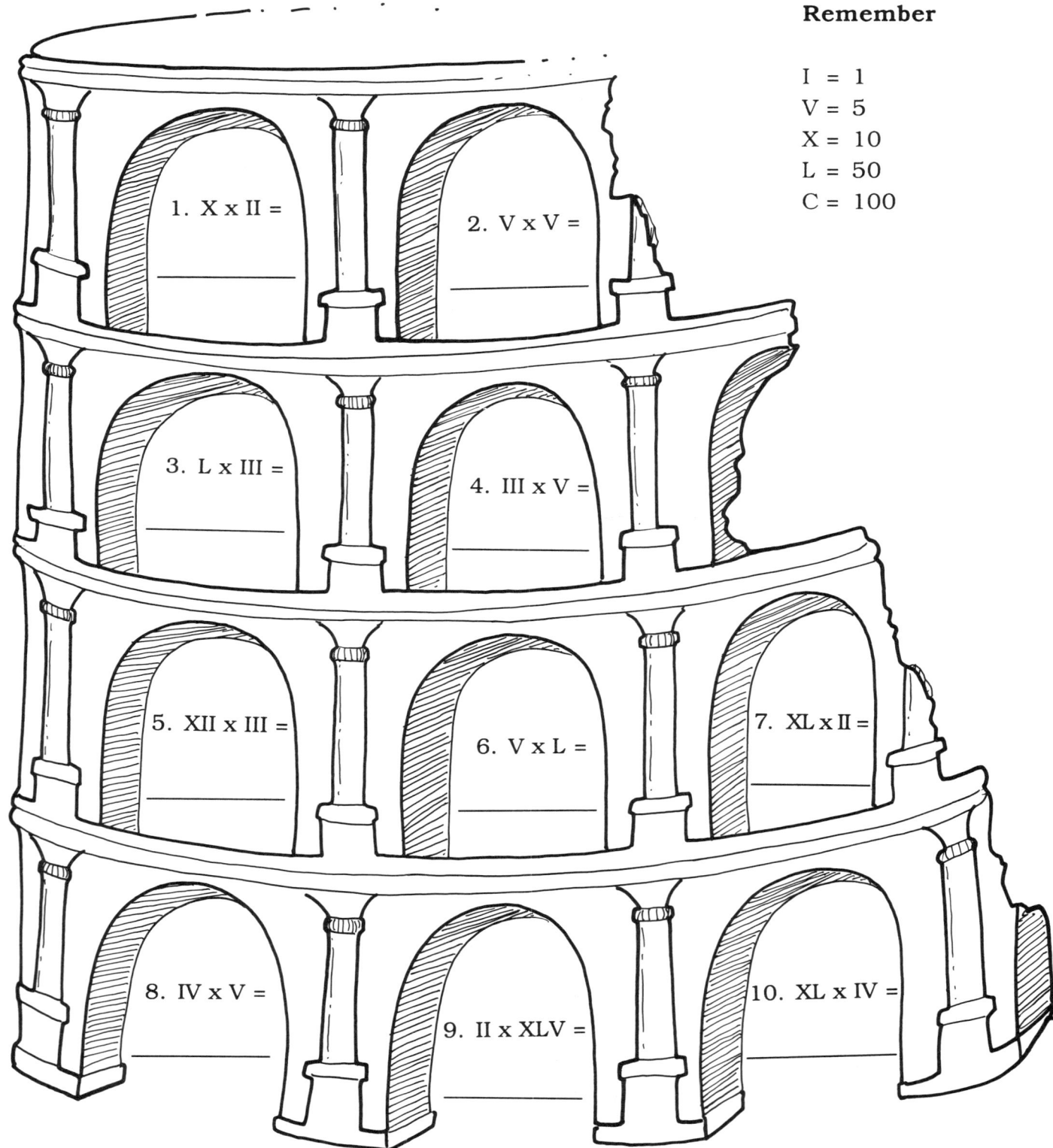

1. X x II = 

2. V x V = 

3. L x III = 

4. III x V = 

5. XII x III = 

6. V x L = 

7. XL x II = 

8. IV x V = 

9. II x XLV = 

10. XL x IV =

Addition of Decimals

# These Are "Sum" Squares!

Can you arrange the numbers on the markers at the bottom of the page so that the four corners of each of three squares add up to 5.9? Move the markers from place to place as you try various solutions. When you've found one that works, write your answers in the circles. (The 1.3 has been placed to help you get started.)

| 1.0 | 1.1 | 1.2 | 1.3 | 1.4 | 1.5 | 1.6 | 1.7 | 1.8 | 1.9 |

Ordering Decimals, Logic,
Subtraction, Averaging

# Start Your Engines!

Here is a list of the twelve fastest drivers during the 1993 Indianapolis 500 time trials. Number them from 1 to 12 with 1 being the fastest and 12 the slowest.

A. _____ Boesel, 222.379 mph

B. _____ Guerrero, 219.645 mph

C. _____ Unser, 221.773 mph

D. _____ Mansell, 220.255 mph

E. _____ Luyendyk, 223.967 mph

F. _____ Johansson, 220.824 mph

G. _____ Sullivan, 219.428 mph

H. _____ Tracy, 220.298 mph

I. _____ Andretti, 223.414 mph

J. _____ Fittipaldi, 220.150 mph

K. _____ Goodyear, 222.344 mph

L. _____ Brayton, 219.637 mph

Junior Racer always studies the list of the top time trial drivers and then picks his favorite to win the big race. Use the clues to find the driver that was Junior's favorite before the 1993 Indy 500.

1. Junior thinks the driver who finished first in the time trials will be overconfident, so he doesn't pick him.

2. Junior knows that in the Indy 500 there are three cars in each row. He knows the drivers that place first, second, and third start in the first row; the fourth, fifth, and sixth place drivers start in the second row; and so on. Junior believes the fourth row is bad luck, so he won't pick his winner from that row.

3. Junior picks his winner from those drivers who are at least 3 miles per hour slower than the first place finisher.

4. Junior never picks a winner whose time trial speed contains the digits 8 or 9.

5. Junior never picks the drivers with either the shortest or the longest last name.

   A. Junior's pick for the winner was _____.

   B. What was the average speed of the top twelve drivers in the 1993 time trials? _____.

   C. Was Junior's pick faster or slower than that average? _____.

   By how much? _____.

**Bonus trivia question: Was Junior's pick for the 1993 Indy 500 correct? ____ If not, could you devise a list of steps that would have led him to the right winner?

Addition, Subtraction, Multiplication, and Division with Decimals; Rounding

# House Hassle

How much did Mr. U.R. Hoam pay for his new house? Follow the directions below to find out. Round each answer to the nearest cent.

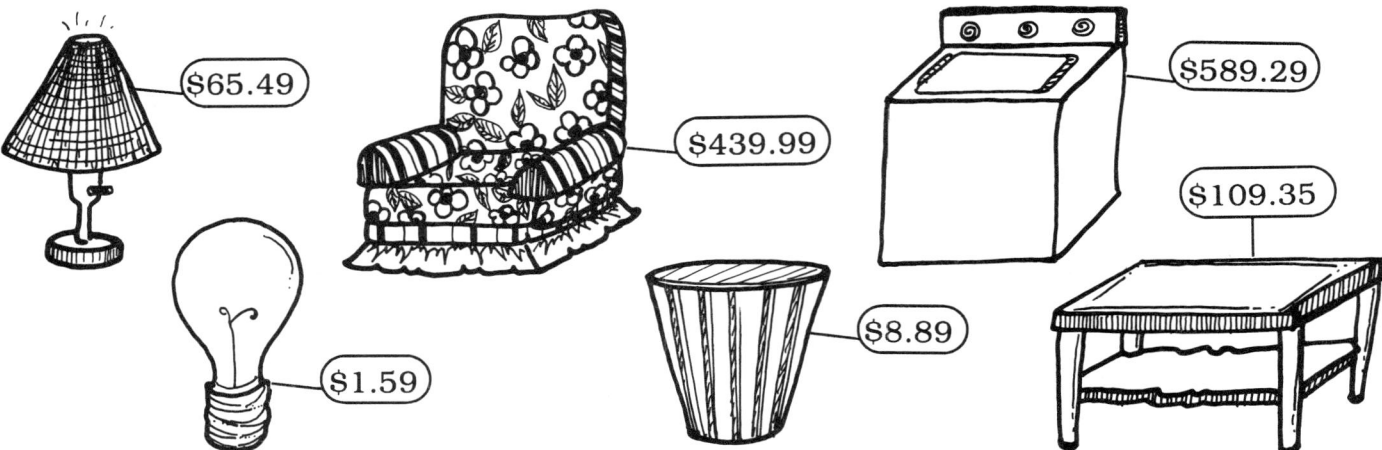

1. Multiply the most expensive item by the least expensive item.

   _____ x _____ = _____

2. Triple your answer.

   _____ x _____ = _____

3. Add to your answer all three prices in the top row.

   _____ + _____ + _____ + _____ = _____

4. Multiply your answer by the price of 2 end tables.

   _____ x _____ = _____

5. Subtract the price of 100 recliners.

   _____ - _____ = _____

6. Finally, to find the final cost of Mr. U.R. Hoam's new house, divide your answer to Step 5 by the cost of a wastebasket.

   _____ ÷ _____ = _____

**Try to write your own sets of computations using the same furnishings that will result in homes that cost approximately $50,000 and $200,000.

Division of Decimals, Patterns

# A"Maze"ing Mathematics II

First find the answer to each problem and write the answer in each box. Then study your answers to find some rule or pattern that will connect Start and Finish. The path must move one box at a time in a vertical or horizontal (not diagonal) direction.

**Start**

| 445.5 ÷ 405 | 13.23 ÷ 6.3 | 204.6 ÷ 66 | 26.01 ÷ 5.1 |
|---|---|---|---|
| 342.22 ÷ 48.2 | 29.766 ÷ 7.26 | 585.6 ÷ 96 | 34.83 ÷ 4.3 |
| 82.901 ÷ 9.11 | 507.6 ÷ 36 | 105.27 ÷ 8.7 | 560.55 ÷ 55.5 |
| 903.9 ÷ 69 | 708.4 ÷ 44 | 246.42 ÷ 22.2 | 211.4 ÷ 14 |
| 564.3 ÷ 33 | 1639.498 ÷ 90.58 | 404.01 ÷ 20.1 | 1301.87 ÷ 61.7 |

**Finish**

# Bug Off!

Why are mosquitoes called "arithmetic bugs"? To find out, number the fractions below from 1 to 13, beginning with the smallest and numbering up to the largest. Then arrange the words under the fractions in the same order, rewriting them in the blanks below. You will then have the answer to the riddle. (The first fraction has been done for you.)

| | | | | | | |
|---|---|---|---|---|---|---|
| | | | | | 1 | |
| $^4/_9$ | $^1/_{10}$ | $^2/_5$ | $^7/_8$ | $^3/_7$ | $^1/_{20}$ | $^1/_2$ |
| divide | add | from | quietly | pleasure | They | your |

| | | | | | |
|---|---|---|---|---|---|
| | | | | | |
| $^1/_4$ | $^{10}/_{18}$ | $^2/_3$ | $^3/_8$ | $^1/_6$ | $^5/_6$ |
| misery | attention | and | subtract | to | multiply |

Answer to riddle:

________ 1 ________ 2 ________ 3 ________ 4 ________ 5 ________ 6

________ 7 ________ 8 ________ 9 ________ 10 ________ 11

________ 12 ________ 13.

Equivalent Fractions,
Ordering Fractions

# A Quest for Success

To uncover an important truth about success, follow the directions at the bottom of the page.

| $3/10$ because | $13/20$ most | $17/20$ are | $11/20$ to, | $4/8$ only |
|---|---|---|---|---|
| $4/5$ they | $8/7$ always | $1/10$ people | $5/20$ money | $5/15$ up |
| $10/9$ plan | $6/8$ every | $6/5$ world | $3/9$ the | $3/20$ may |
| $7/4$ can | $2/5$ are | $7/10$ succeed | $3/5$ but | $4/12$ will |
| $1/20$ some | $25/100$ finish | $3/4$ because | $11/10$ wealth | $19/20$ to |
| $9/10$ determined | $6/12$ driven | $10/20$ won't | $1/5$ succeed | $2/8$ win |
| $2/10$ play | $7/20$ they | $5/10$ start | $3/12$ record | $9/20$ destined |

1. When you cross out a fraction in the chart above, also cross out the word with it.
2. Cross out all fractions greater than 1.
3. Cross out all fractions that are equivalent to $1/4$ and $1/2$.
4. Cross out any other fractions that are not in lowest terms.
5. Circle the remaining fractions. Arrange them in order from least to greatest here. Then write the corresponding words in order to form a message about success. _____

Spatial Relations,
Equivalent Fractions

# Fractured Cubes

In each row, the same cube is shown in three different positions. Your job is to figure out which fraction should replace the question mark on the last cube in each row.

**Hint:** Opposite faces contain equivalent fractions.

1.

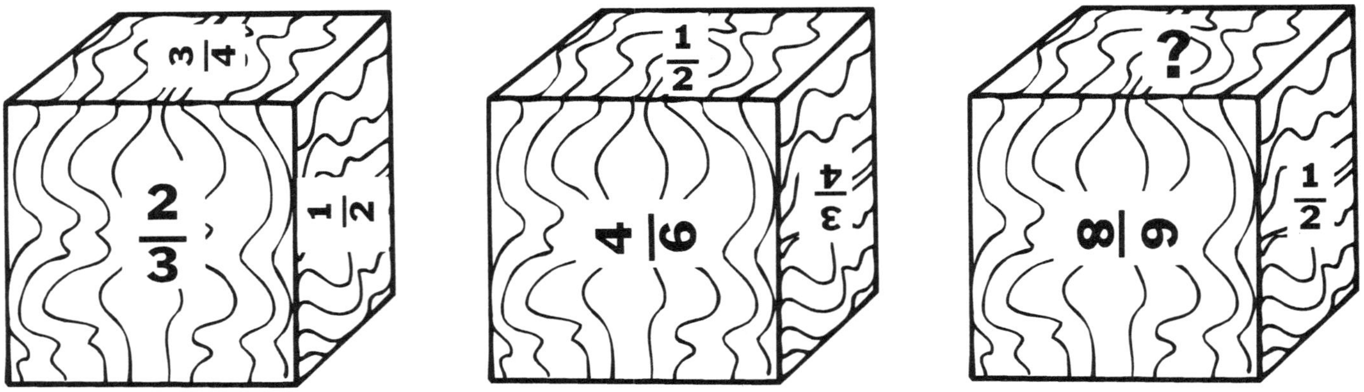

Answer: _____

2.

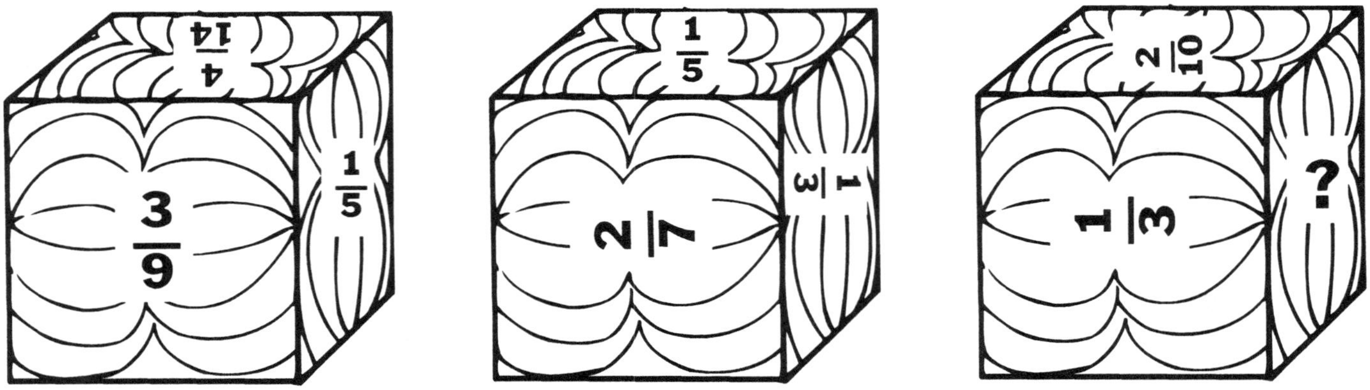

Answer: _____

3.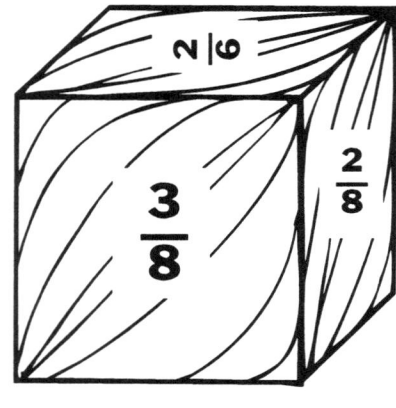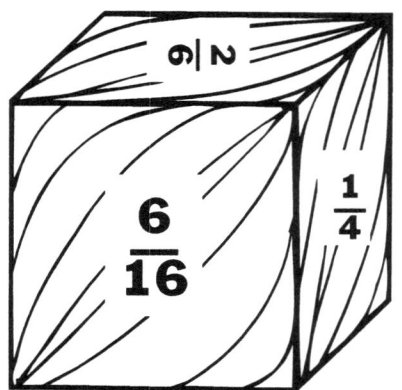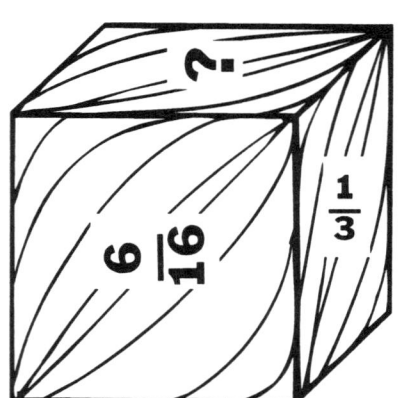

Answer: _____

Addition and Subtraction of Fractions with Unlike Denominators, Patterns

# A"Maze"ing Mathematics III

First find the answer to each problem and write it in each box. (Reduce your answers to mixed numerals in lowest terms.) Then study your answers and find some rule or pattern that will connect Start and Finish. The path must move one box at a time in a vertical or horizontal (not diagonal) direction.

**Start**

| $1\tfrac{5}{8} + \tfrac{5}{8} + \tfrac{3}{4}$ | $7 - 3\tfrac{3}{4}$ | $1\tfrac{5}{6} + 1\tfrac{2}{3}$ | $2\tfrac{7}{12} - 1\tfrac{5}{12}$ |
|---|---|---|---|
| $\tfrac{3}{8} + 1\tfrac{1}{4} + 1\tfrac{1}{8}$ | $5 - 2\tfrac{7}{8}$ | $4\tfrac{2}{3} - 2\tfrac{11}{12}$ | $3\tfrac{1}{6} - 1\tfrac{2}{3}$ |
| $1\tfrac{5}{6} + \tfrac{2}{3}$ | $3\tfrac{7}{8} - 1\tfrac{5}{8}$ | $\tfrac{7}{8} + \tfrac{3}{4} + \tfrac{3}{8}$ | $\tfrac{13}{20} + \tfrac{9}{20} + \tfrac{3}{20}$ |
| $2\tfrac{4}{5} - 1\tfrac{3}{10}$ | $\tfrac{5}{8} + \tfrac{1}{2}$ | $1\tfrac{7}{16} - \tfrac{13}{16}$ | $\tfrac{12}{16} + \tfrac{1}{4}$ |
| $3 - 2\tfrac{1}{10}$ | $1\tfrac{7}{12} - \tfrac{3}{4}$ | $\tfrac{1}{2} + \tfrac{11}{12} + \tfrac{7}{12}$ | $3\tfrac{5}{8} - 2\tfrac{7}{8}$ |

**Finish**

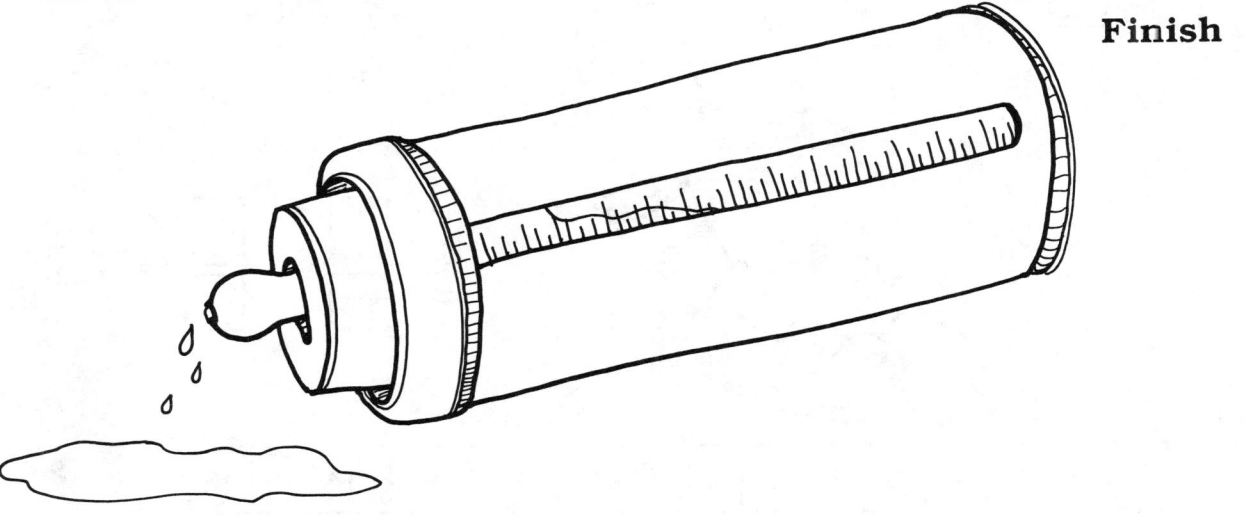

47

Copyright © 1994, Good Apple

GA1505

Addition and Subtraction of Fractions with Unlike Denominators

# Once Is Enough

Use the numbers from 1 to 16 to complete the following problems. You may use each number only once. Use the markers at the bottom of the page to try various solutions. When you're done, write the numbers in the boxes. The 1, 7, and 16 have been placed to help you get started. (Answers to problems have been written in simplest form.)

A. $\dfrac{7}{\Box} + \dfrac{\Box}{\Box} = 1$

B. $\dfrac{\Box}{\Box} + \dfrac{\Box}{\Box} = 1\,{}^{4}/_{15}$

C. $\dfrac{\Box}{16} - \dfrac{\Box}{\Box} = {}^{3}/_{16}$

D. $\dfrac{1}{\Box} - \dfrac{\Box}{\Box} = {}^{1}/_{22}$

| 1 | 2 | 3 | 4 | 5 | 6 | 7 | 8 | 9 | 10 | 11 | 12 | 13 | 14 | 15 | 16 |

Multiplication and Division of Fractions

# Once Again

Use the numbers from 1 to 16 to complete the following problems. You may use each number only *once*. Use the markers at the bottom of the page to try various solutions. When you're done, write the numbers in the boxes. The 1, 10, and 16 have been placed to help you get started. (Answers to problems have been written in simplest form.)

A. $\dfrac{\square}{10} \times \dfrac{\square}{\square} = \dfrac{3}{40}$

B. $\dfrac{\square}{\square} \div \dfrac{\square}{\square} = 1\dfrac{6}{7}$

C. $\dfrac{\square}{16} \div \dfrac{1}{\square} = 1\dfrac{3}{4}$

D. $\dfrac{\square}{\square} \times \dfrac{\square}{\square} = \dfrac{11}{27}$

| 1 | 2 | 3 | 4 | 5 | 6 | 7 | 8 | 9 | 10 | 11 | 12 | 13 | 14 | 15 | 16 |

Addition and Multiplication of Fractions, Number Patterns

# "Sum"Thing's Strange!

Some pairs of numbers have sums which are equal to their products. Here are two examples.

$2 + 2 = 4$      $1\frac{1}{3} + 4 = 5\frac{1}{3}$

$2 \times 2 = 4$      $1\frac{1}{3} \times 4 = 5\frac{1}{3}$

Now look at these equations. First, try to predict which pairs will yield equal sums and products. Circle the letter in front of the pairs that you think will work. Then do the computations and write your answers in the blanks. Were you right?

A. $1\frac{1}{5} + 6 =$ _____

   $1\frac{1}{5} + 6 =$ _____

B. $2\frac{1}{2} + 2 =$ _____

   $2\frac{1}{2} \times 2 =$ _____

C. $1\frac{1}{3} + 3 =$ _____

   $1\frac{1}{3} \times 3 =$ _____

D. $1\frac{1}{2} + 3 =$ _____

   $1\frac{1}{2} \times 3 =$ _____

E. $2\frac{1}{3} + 1 =$ _____

   $2\frac{1}{3} \times 1 =$ _____

F. $1\frac{1}{6} + 7 =$ _____

   $1\frac{1}{6} \times 7 =$ _____

**Now try to come up with number pairs of your own that will yield equal sums and products.

Estimating Percentages

# Percentage Predictions

How well can you predict the outcome of these percentage problems?
* 40% of 30        *55% of 80

By using some "mental math" you can arrive at the exact answer or a reasonable estimate rather easily. For example, to find 40% of 30 you can simply multiply 4 x 3 mentally and arrive at 12. To find 55% of 80, you can reason that since 50% of 80 would be 40, then 55% must be a little larger than 40. (The actual answer is 44.) Using similar methods, try to match each problem below to a reasonable answer. You will have four extra answers. If your work is correct, you will be able to rearrange the letters in front of the extra answers and spell a word that will let you know how you did.

1. _____ 40% of 200
2. _____ 90% of 150
3. _____ 30% of 210
4. _____ 5% of 120
5. _____ 60% of 90
6. _____ 20% of 65
7. _____ 80% of 200
8. _____ 10% of 90
9. _____ 75% of 160
10. _____ 30% of 80
11. _____ 45% of 200
12. _____ 17% of 300
13. _____ 70% of 60
14. _____ 36% of 50

A. 200
B. 6
D. 160
F. 18
G. 80
H. 24
J. 54
K. 2
M. 9
O. 70
P. 135
Q. 42
S. 13
T. 90
V. 63
W. 51
Y. 30
Z. 120

Extra Letters: __ __ __ __        Unscrambled: __ __ __ __

Computations with Percent and Money

# Clancy's Clothing Sale

Some of Clancy's sale prices are genuine bargains; others are inaccurate tricks. Circle the parts of each advertisement that are inaccurate or inconsistent. In the space provided, tell what's wrong with the ad and show one way it could be corrected. If the ad is correct as written, simply write *okay* in the work space. If there's not enough information to tell if the ad is correct, write *need more info* in the space.

| Advertisement | Work Space |
|---|---|
| **A. Kid's Sneakers** — Save 35%  Were $19.95    Now just $12.95 | |
| **B. Men's Silk Ties** — 25% off  Were $12 each    Now just 3 for $30 | |
| **C. Ladies' Dresses** — 30%-50% off  Were $49.99-69.99  Now all are on sale for only $39.99! | |

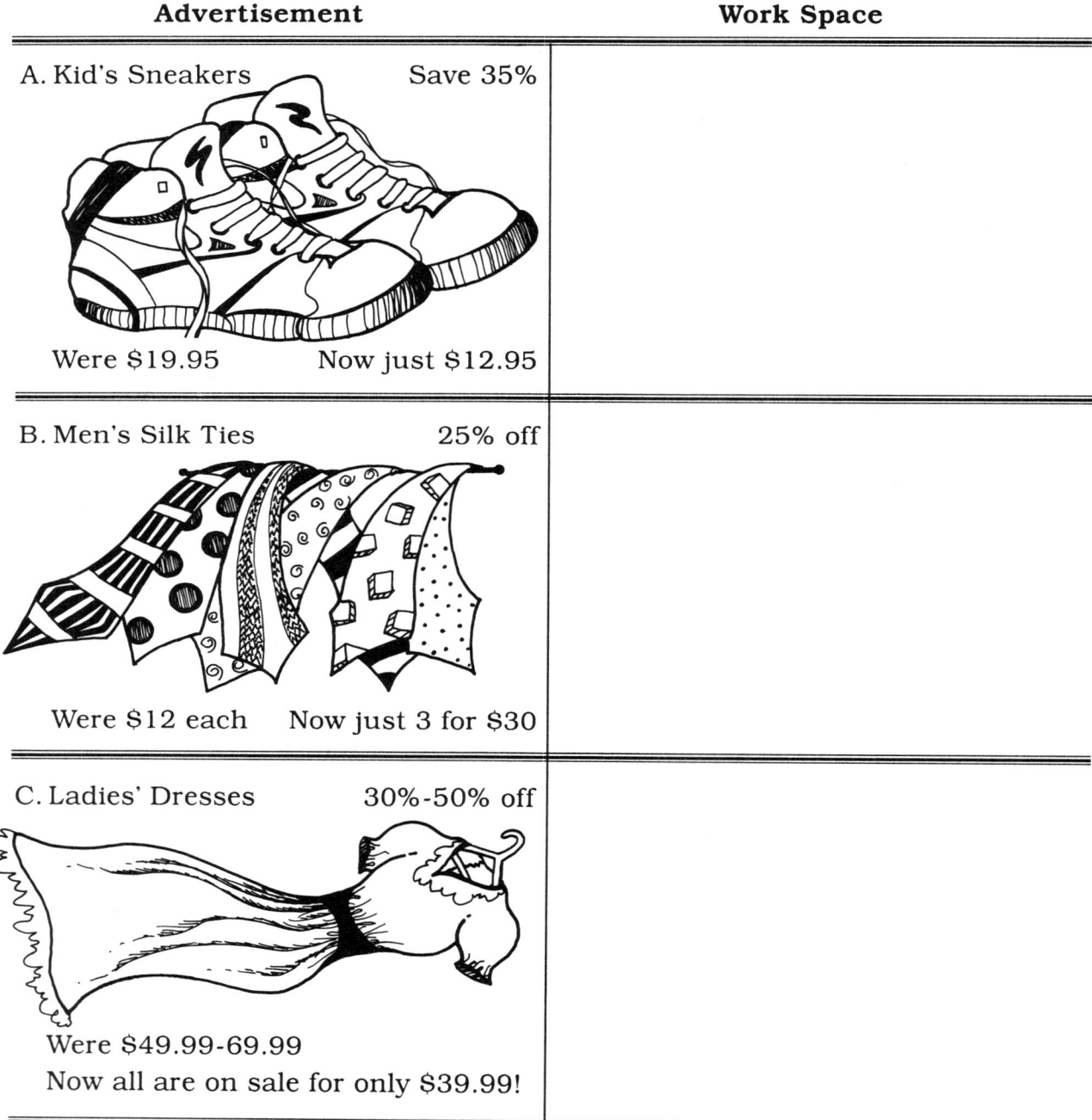

| Advertisement | Work Space |
|---|---|
| D. Men's Suits — Save over 20%  Were $250–$400   Now $195–$310 | |
| E. Kids' Jeans — Save up to 60% 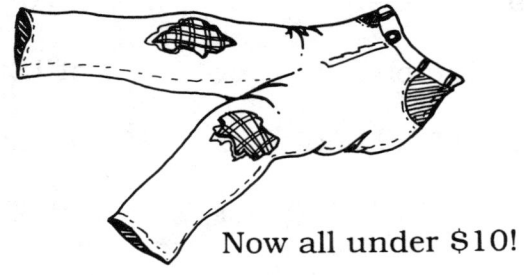 Now all under $10! Were $12.99–$21.99 | |
| F. Take 25% off all socks  Were $1.99–$8.99 Now $1.69–$7.64 | |
| G. Kids' Raincoats — Save at least 25% 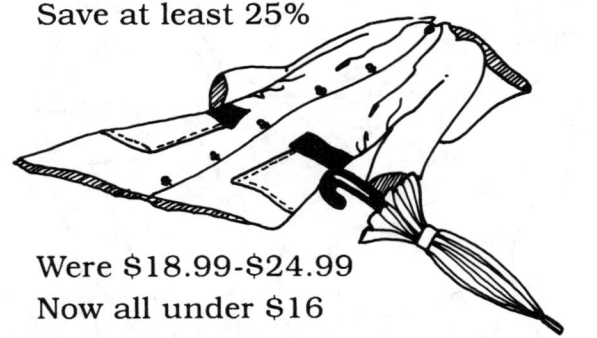 Were $18.99–$24.99 Now all under $16 | |

**Now write your own ads for a classmate to check.

Finding Percentages and
Fractional Parts of Whole Numbers,
Estimation

# Score More

Circle the item in each row that's worth more. If they are worth the same, circle *both*. See how many you can solve in your head.

1. 30% of 90   or   60% of 30
2. 25% of 20   or   20% of 25
3. 3/5 of 20   or   5/8 of 24
4. 10% of 90   or   3% of 300
5. 3/4 of 64   or   4/5 of 50
6. 24% of 200   or   12% of 400
7. 1/2 of 144   or   9/10 of 90
8. 75% of 200   or   60% of 300
9. 5% of 90   or   6% of 70
10. 1/8 of 160   or   1/5 of 120
11. 18% of 300   or   30% of 180
12. 15% of 2000   or   10% of 3200

Finding Possible Solutions,
Computations with Money and Percent

# Sports Court Sale

The Sports Court sporting goods store has posted this sign to advertise its current sale:

15% off all footwear
25% off all balls
(5% Sales tax added to all purchases.)

Special Sale!

Use the store's regular price list and the information on the sale sign to answer the questions below. (Round to the nearest cent.)

### Regular Price List

Soccer ball–$16
Baseball–$7
Football–$25
Basketball–$32

Golf clubs–$98
Tennis racquets–$39
Baseball bat–$15
Hockey stick–$29

Baseball glove–$19
Roller skates–$49
Baseball shoes–$34

1. What is the final cost of a baseball, bat, glove, and shoes during the sale?

   _____

2. Jessica received $50 for her birthday and wants to spend it at Sports Court. If she buys two items, receives some change, and does not purchase any baseball equipment, what are her choices? How much will each cost?

   **Choice**                                   **Cost**

   _____          _____

   _____          _____

   _____          _____

   _____          _____

3. Suppose you have $100 to spend at this Sports Court. What would you buy?

   _____

   What would be your total cost? _____

Computing Interest

# Saver's Savor

You have carefully saved $1000 in extra cash. Three of your friends who haven't been so careful all want to borrow money from you. Each friend offers a different deal. Decide how much interest you will earn under each plan.

Joe wants to borrow $700 for 6 months at 8% per year.

  A. Interest you would earn: _____

Jill wants to borrow $900 for 1 year at 5% per year

  B. Interest you would earn: _____

Jim wants to borrow $800 for 9 months at 7% per year

  C. Interest you would earn: _____

  D. Where would you earn the most interest? _____

Now suppose that for each plan above you invest any part of the $1000 that isn't loaned out. Your investment pays a rate of 6% per year. Compute the investment interest for each of the three plans. Under Joe's plan, for example, you need to figure $300 at 6% for the full year, plus $700 at 6% for 6 months.

               Interest on Investment:

  E. Joe's plan:          _____

  F. Jill's plan:         _____

  G. Jim's plan:        _____

  H. Now add the total interest (interest paid by each friend plus interest on your investment) under each plan. Which plan now offers the most profitable package? _____

# Surprise Graph I

Graphing Coordinate Pairs

Draw a surprise picture by making and connecting dots on the graph on page 58. Beginning with the upper left, move down the columns, finding each coordinate pair's point on the graph and make a dot on it. Then find the next dot and connect it to the first dot you drew. (If the directions say "lift pencil," *don't* connect with the previous dot.) It may help to cross out each coordinate pair as you plot it on the graph.

| | | | | |
|---|---|---|---|---|
| (1, 0) | (16, 13) | (3, 5) | lift pencil | lift pencil |
| (1, 7) | (16, 10) | (5, 6) | (17, 2) | (9, 7) |
| (3, 7) | (18, 10) | (7, 5) | (17, 5) | (9, 9) |
| (3, 11) | (18, 15) | (7, 2) | (19, 6) | (12, 11) |
| (2, 11) | (17, 15) | (3, 2) | (21, 5) | (15, 9) |
| (4, 14) | (19, 18) | | (21, 2) | (15, 7) |
| (6, 11) | (21, 15) | | (17, 2) | (9, 7) |
| (5, 11) | (20, 15) | | lift pencil | lift pencil |
| (5, 7) | (20, 10) | | (9, 0) | (13, 13) |
| (8, 7) | (21, 10) | | (9, 6) | (11, 13) |
| (8, 13) | (21, 7) | | (15, 6) | (11, 18) |
| (10, 13) | (23, 7) | | (15, 0) | (13, 18) |
| (10, 18) | (23, 0) | | lift pencil | (13, 13) |
| (9, 18) | lift pencil | | (12, 0) | lift pencil |
| (12, 21) | (3, 2) | | (12, 6) | (12, 21) |
| (15, 18) | | | | (12, 27) |
| (14, 18) | | | | (7, 24) |
| (14, 13) | | | | (12, 24) |

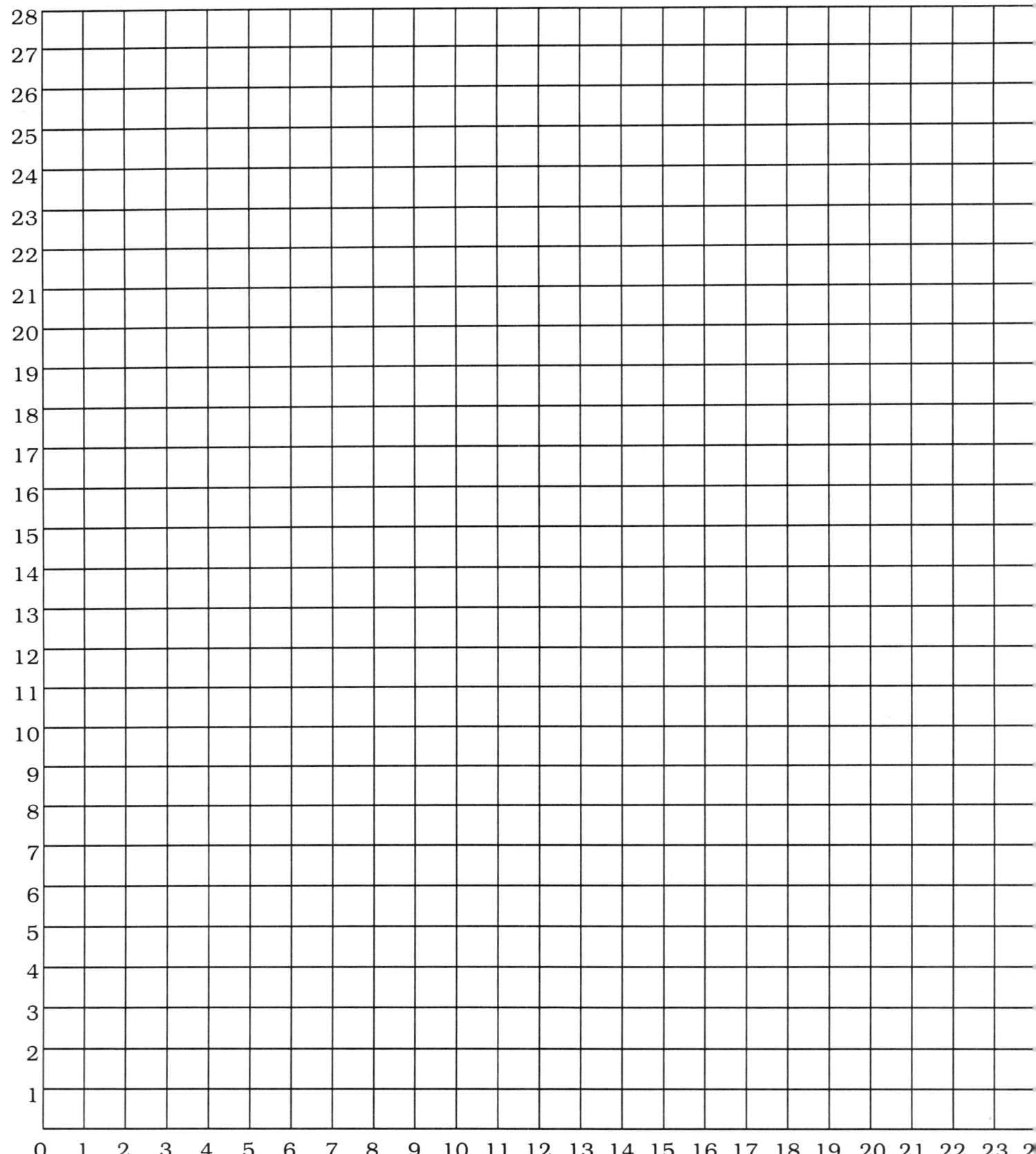

# Surprise Graph II

Graphing Coordinate Pairs

Using the graph on page 58, draw a surprise picture by making and connecting dots on the graph. For each coordinate pair, find that point on the graph and make a dot on it. Then find the next dot and connect it to the first dot you drew. (If the directions say "lift pencil," *don't* connect with the previous dot.) It may help to cross out each coordinate pair as you plot it on the graph.

| | | | | |
|---|---|---|---|---|
| (1, 8) | (14, 24) | (9, 17) | (12, 18) | (17, 9) |
| (1, 1) | (13, 23) | (6, 17) | (9, 12) | (19, 11) |
| (10, 2) | (12, 24) | (7, 19) | (9, 11) | (21, 11) |
| (14, 2) | (11, 23) | (8, 19) | (13, 11) | (21, 9) |
| (23, 1) | (10, 24) | (9, 17) | lift pencil | (17, 6) |
| (23, 8) | (8, 23) | lift pencil | (3, 18) | (7, 6) |
| (14, 4) | (8, 24) | (15, 17) | (2, 19) | (3, 9) |
| (10, 4) | (5, 22) | (14, 19) | (1, 18) | (3, 11) |
| (1, 8) | (5, 24) | (15, 21) | (1, 14) | (5, 11) |
| lift pencil | (2, 23) | (18, 21) | (3, 12) | (7, 9) |
| (8, 27) | (8, 27) | (19, 19) | lift pencil | (17, 9) |
| (16, 27) | lift pencil | (18, 17) | (21, 18) | lift pencil |
| (22, 23) | (6, 17) | (15, 17) | (22, 19) | (4, 10) |
| (19, 24) | (5, 19) | (16, 19) | (23, 18) | (7, 8) |
| (19, 22) | (6, 21) | (17, 19) | (23, 14) | (17, 8) |
| (16, 24) | (9, 21) | (18, 17) | (21, 12) | (20, 10) |
| (16, 23) | (10, 19) | lift pencil | lift pencil | |

Using Coordinates,
Making a Scale Drawing,
Following Directions

# Mapmaker Mo-o-o-ves

Using the graph on page 58, complete a map of Cow County. The top of the page is north. Use a scale of 4 squares: 10 miles. Label these locations.

1. Udderville is located at point (6, 4).
2. Jersey City is 25 miles straight north of Udderville.
3. Cudtown is 15 miles east and 20 miles north of Jersey City.
4. Draw a pasture that is 10 miles square. Its southwestern corner is 15 miles straight east of Cudtown.
5. Draw another pasture that is 25 miles from east to west and 10 miles from north to south. Its northeastern corner is 40 miles east of Jersey City.
6. Beefsteak is 10 miles south of the southeastern corner of the larger pasture.
7. Guernsey is 10 miles west and 5 miles north of Cudville.
8. Calfton is 10 miles north and 15 miles east of Udderville.
9. Roads have been built along the shortest route between these cities: Cudtown and Guernsey, Guernsey and Jersey City, Jersey City and Udderville, Udderville and Beefsteak, Beefsteak and Calfton, Calfton and Jersey City, Calfton and Udderville, Jersey City and Cudtown.
    A. Which two cities are connected by the longest road?
    _____ and _____
    B. How long (in miles) is the longest road? _____

**Add other towns, lakes, roads, or other features to your map. Write directions so that a classmate could add the same things to his or her map.

Using Coordinates in 4 Quadrants,
Making a Scale Drawing

# Putter's Plotting

Pete is planning a nine-hole miniature golf course. Use the graph on page 62 to help Pete plot the holes. From the clues below, label each point where a hole belongs.

A. Hole 1 is at point (5, -3).
B. Hole 2 is 8 yards east and 24 yards south of Hole 1.
C. Hole 3 is 40 yards due west of Hole 2.
D. Hole 4 is at point (-3, 5).
E. Hole 5 is 32 yards south and 24 yards west of Hole 4.
F. Hole 6 is 4 yards west and 24 yards north of Hole 6.
G. Hole 7 is at point (-8, 7).
H. Hole 8 is 4 yards north and 40 yards east of Hole 7.

**Pete will put a windmill near one of the holes. Can you figure out from these clues where it will go?
• At least one of the hole's coordinates is negative.
• It is an even-numbered hole closest to the center of the graph.

Draw a windmill by the correct hole on your graph.

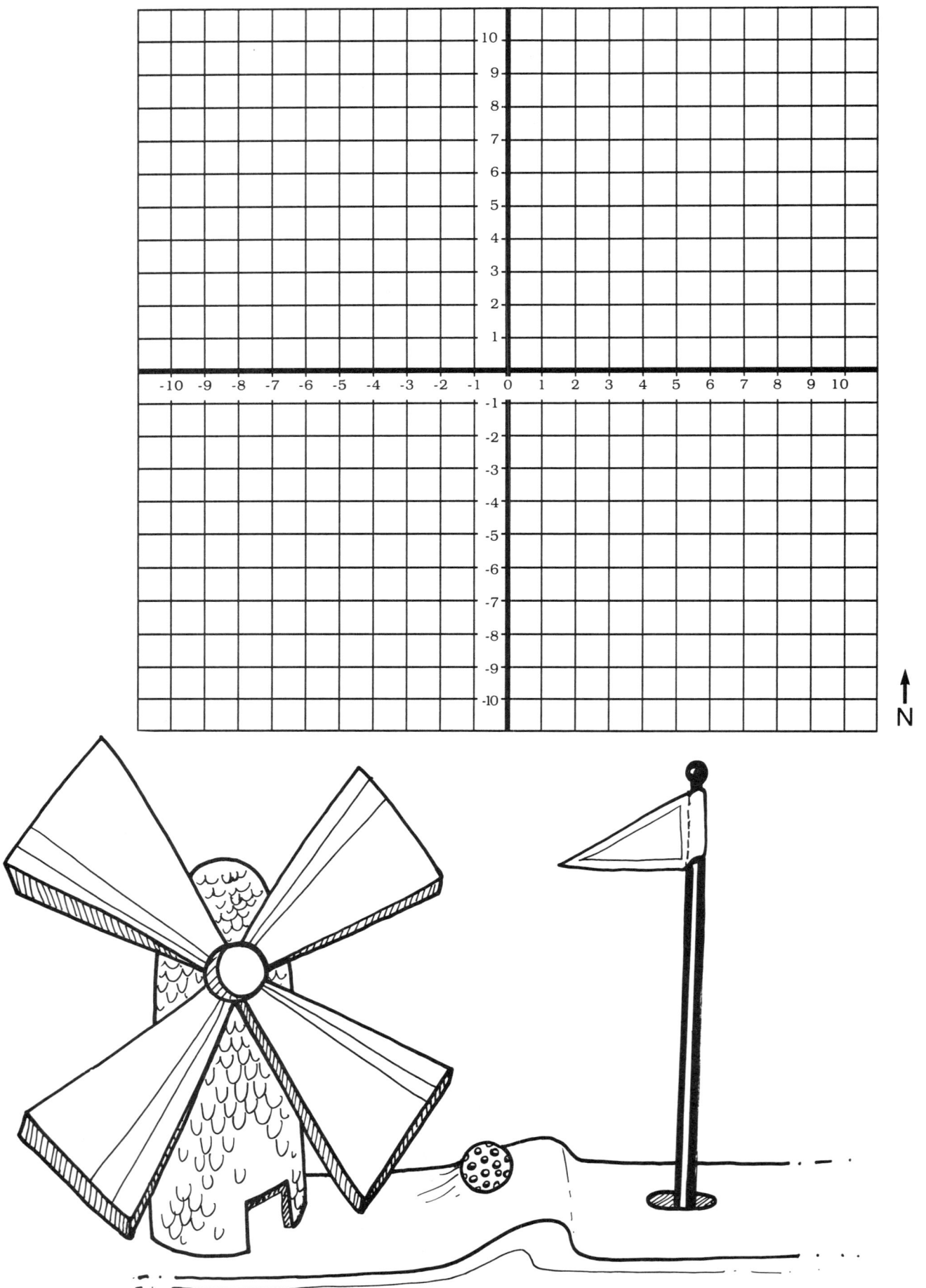

Completing a Chart,
Trying Possible Combinations,
Adding Cents

# Sweet Tooth

Betsy spent exactly $1.00 at the candy store and bought exactly 50 pieces of candy. She bought only gumdrops, lollipops, and candy canes. Can you figure out how many pieces of each kind of candy she bought?

Gumdrops
2 for 1¢

Lollipops
10¢

Candy Canes
5¢

(This type of problem requires some trial-and-error work. Use this statistical chart to keep track of your guesses.)

|   | Number of Gumdrops | Cost of Gumdrops | Number of Lollipops | Cost of Gumdrops | Number of Candy Canes | Cost of Candy Canes | Total Number of Pieces | Total Cost |
|---|---|---|---|---|---|---|---|---|
| 1 | | | | | | | | |
| 2 | | | | | | | | |
| 3 | | | | | | | | |
| 4 | | | | | | | | |
| | | | | | | | | |
| | | | | | | | | |
| | | | | | | | | |
| | | | | | | | | |

Answer:
Betsy bought _____ gumdrops, _____ lollipops, and _____ candy canes.

Compiling Statistics,
Finding Averages,
Making Predictions

# Paper Clip Ponderings

A. On another sheet of paper, draw a circle with a 4-inch diameter. Put ten paper clips in your hand, hold them about 6 inches above the circle and drop them all at once. How many clips landed *completely* inside the circle? _____ Repeat this until you've had a total of ten drops. Write the result of each drop here. ____ ____ ____ ____ ____ ____ ____ ____ ____
Now compute the average number of clips that landed completely inside the circle. Average for a 4-inch circle. _____

B. Now make a prediction. Would you expect your average to go up or down as you *increase* the size of the circle? _____

C. Check it out. Draw a circle with a 6-inch diameter. Do ten drops. Write the number of clips completely inside the circle for each drop here.

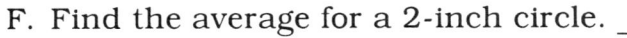

Find the average for a 6-inch circle. _____

D. Make another prediction. Would you expect your average to go up or down as you decrease the size of the circle? _____

E. Check your prediction. Draw a circle with a 2-inch diameter. Write the results of ten drops here.

____ ____ ____ ____ ____ ____ ____ ____ ____ ____

F. Find the average for a 2-inch circle. _____

Completing a Line Graph

Use the information you recorded on page 64 to complete this line graph.

A. Write a title for your graph.
B. Record the results of the ten drops with the 4-inch circle in red.
C. Record the results of the drops with the 6-inch circle in blue.
D. Record the results of the drops with the 2-inch circle in black.

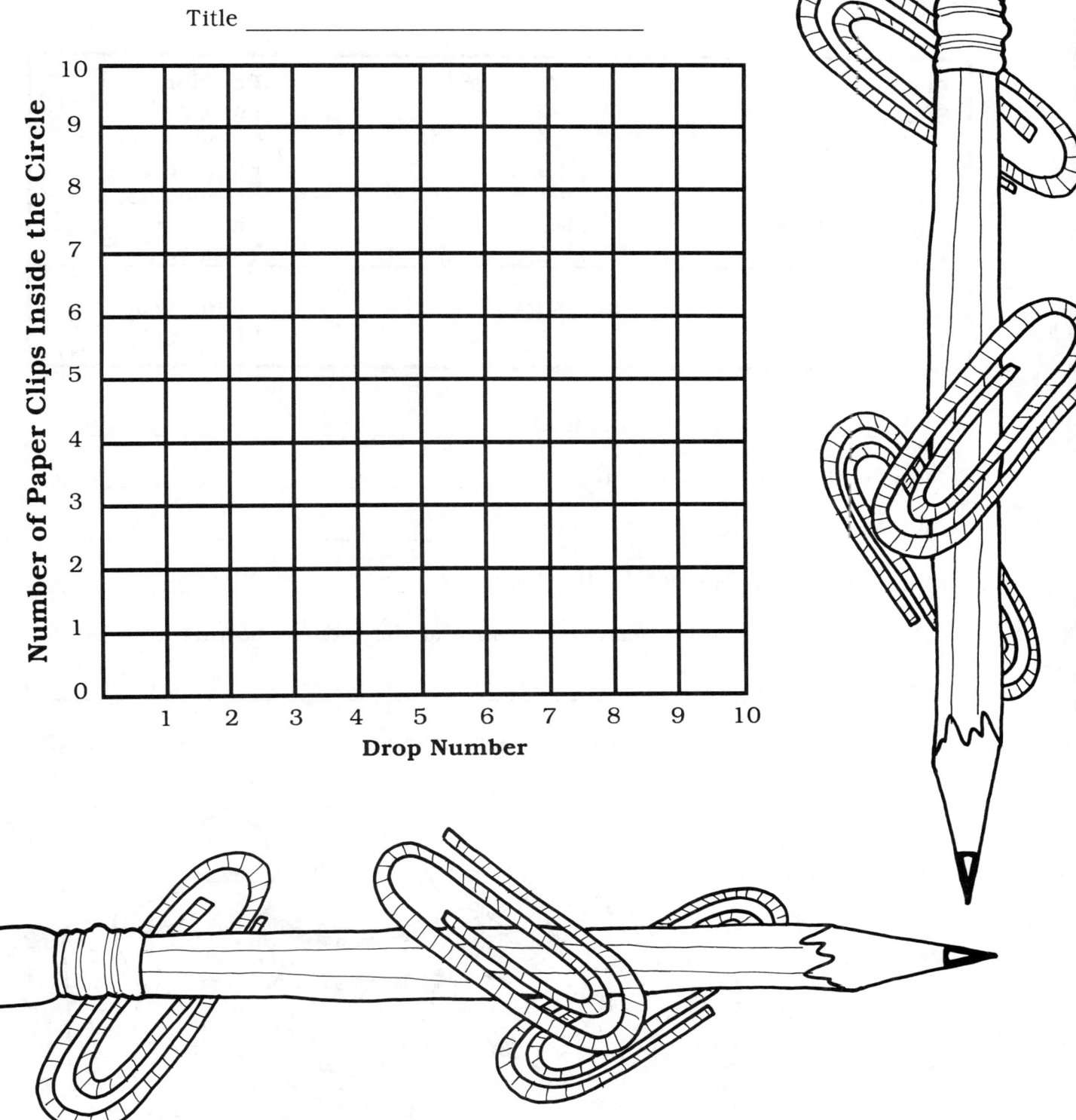

Interpreting Information in a Chart,
Finding the Median and the Mean

# Batter Up!

Part One:

Coach Rookie is getting his baseball team organized for opening day. Here is a roster of his nine top players, along with their preseason batting averages and positions.

| Player | Batting Average | Position | |
|---|---|---|---|
| Samuel Strikeout | .092 | Pitcher | |
| Mickey Mitt | .218 | Catcher | Infield |
| Fred First | .199 | First base | |
| Danny Double | .361 | Second base | |
| Shorty McAllister | .387 | Shortstop | |
| Tommy Triple | .265 | Third base | |
| Orville Left | .382 | Left field | |
| Guard D. Fence | .284 | Center field | Outfield |
| Wilbur Right | .236 | Right field | |

1. Who is the best batter on the team? _____

2. Who is the worst batter on the team? _____

3. Whose batting average is the median? _____

4. What is the team's mean batting average? _____

5. Find the difference between the mean and the median batting averages for the team. _____

6. Now drop the pitcher's batting average and find the mean for the rest of the team. _____

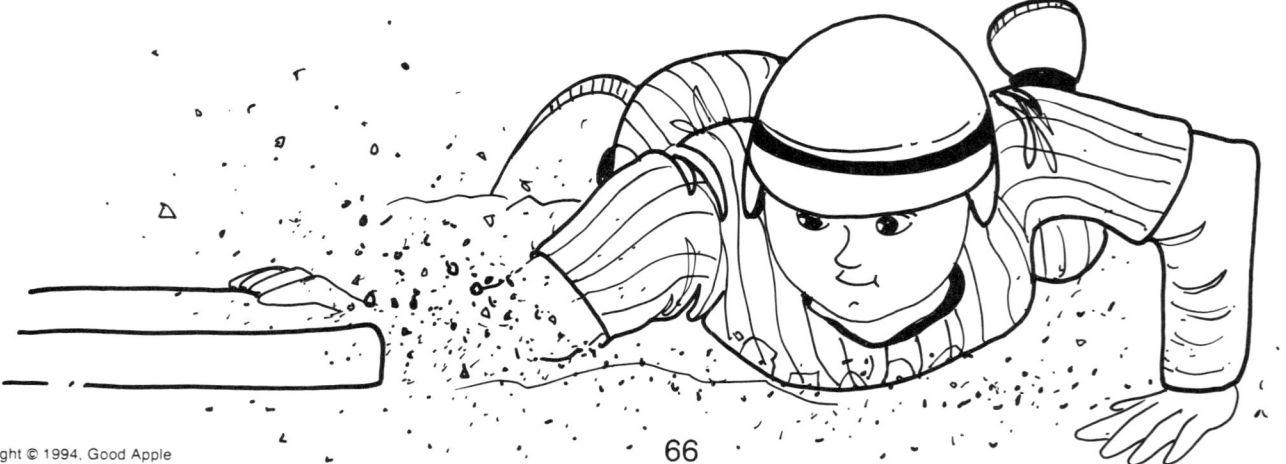

Finding the Number of Possible Outcomes

# Batter Up!

Part Two:
Coach Rookie is experimenting with different batting orders.

A. If the coach changes the lineup of his nine batters every game, how many games would his team have to play in order to try every possible lineup?

___

B. Suppose that Coach Rookie's team played one game every day, all year long. How many years would it take to play all the games in step A?

___

C. Coach Rookie comes to his senses and realizes he has to limit the number of lineups he's willing to try. He decides to put his best batter, Shorty McAllister, first and worst batter, Samuel Strikeout, last. Now how many different lineups can he try?

___

D. There are still too many choices. Coach Rookie decides to start with Shorty McAllister, then go to his three outfielders, then move to the rest of the infield, and then finish each time with the pitcher, Samuel Strikeout. Now how many options does the coach have?

___

Finding the Probability of an Outcome, Using Fractions

# April Odds

Using the calendar page shown here, find the probability of each event listed below. Write your answer as a fraction in lowest terms. For example, if the chances of an event happening were 2 out of 30, it would be written as 2/30 and then reduced to 1/15.

If you were to close your eyes and randomly point to a date on the calendar below, what is the probabiity that you would choose:

A. an even-numbered date? _____

B. an odd-numbered date? _____

C. a Sunday? _____

D. a Saturday? _____

E. a weekday? _____

F. a date containing a 2? _____

G. a date containing two 3's? _____

| **April** | | | | | | |
|---|---|---|---|---|---|---|
| S | M | T | W | Th | F | S |
|   |   |   |   |   | 1 | 2 |
| 3 | 4 | 5 | 6 | 7 | 8 | 9 |
| 10 | 11 | 12 | 13 | 14 | 15 | 16 |
| 17 | 18 | 19 | 20 | 21 | 22 | 23 |
| 24 | 25 | 26 | 27 | 28 | 29 | 30 |

**\*\*Write two questions of your own about probability and the calendar for a classmate to answer.

Finding Possible Outcomes,
Finding Probability

## The New Shoes Blues

Sam Wingtip, a hardworking shoe salesman, has had a rough day. He's been waiting on customers for many hours but has made very few sales. His last customer of the day, Miss Persnickety, just tried on five pairs of shoes. Each pair was a different color: red, black, green, yellow, or white. Sam has already placed the left shoe of each pair in the correct box. Now, tired and discouraged, he randomly picks up each right shoe and tosses it into one of the boxes.

A. In how many ways could the shoes be paired? _____

B. What is the probability that Sam randomly placed all five shoes in the correct box?

_____

C. What is the probability that Sam placed exactly four right shoes into the correct boxes?

_____

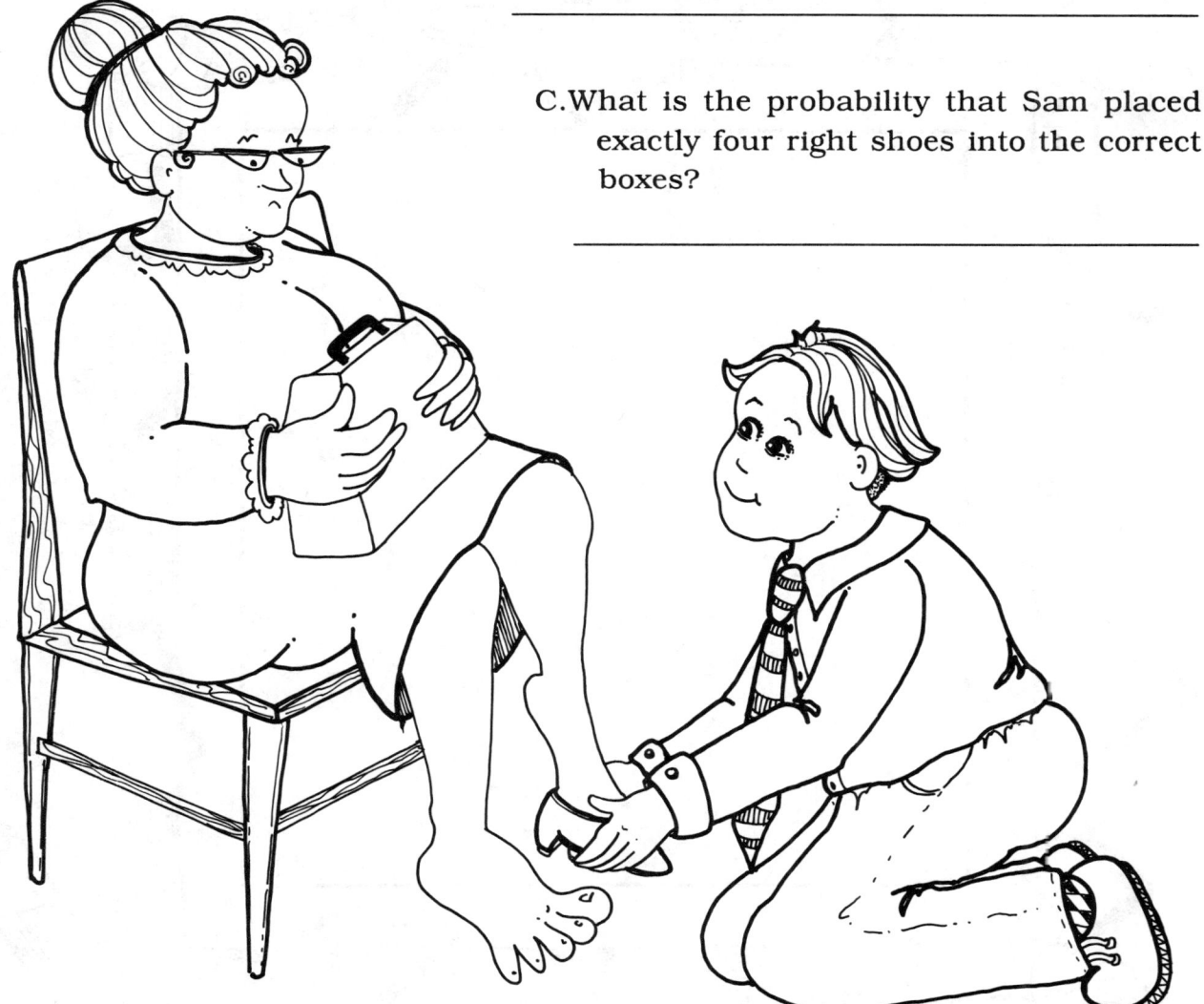

Constructing Regions with
Intersecting Lines

# Square Dare

Divide this square into as many sections as possible. Follow these rules:
1. Draw exactly four straight lines.
2. Your lines may intersect one another.
3. Make sure each line connects two sides of the square.

(Hint: Use four straws or narrow strips of paper and move them around as you try various solutions. This saves a lot of erasing!) When you think you've found the largest number of sections possible, draw your lines into the square and number your sections.

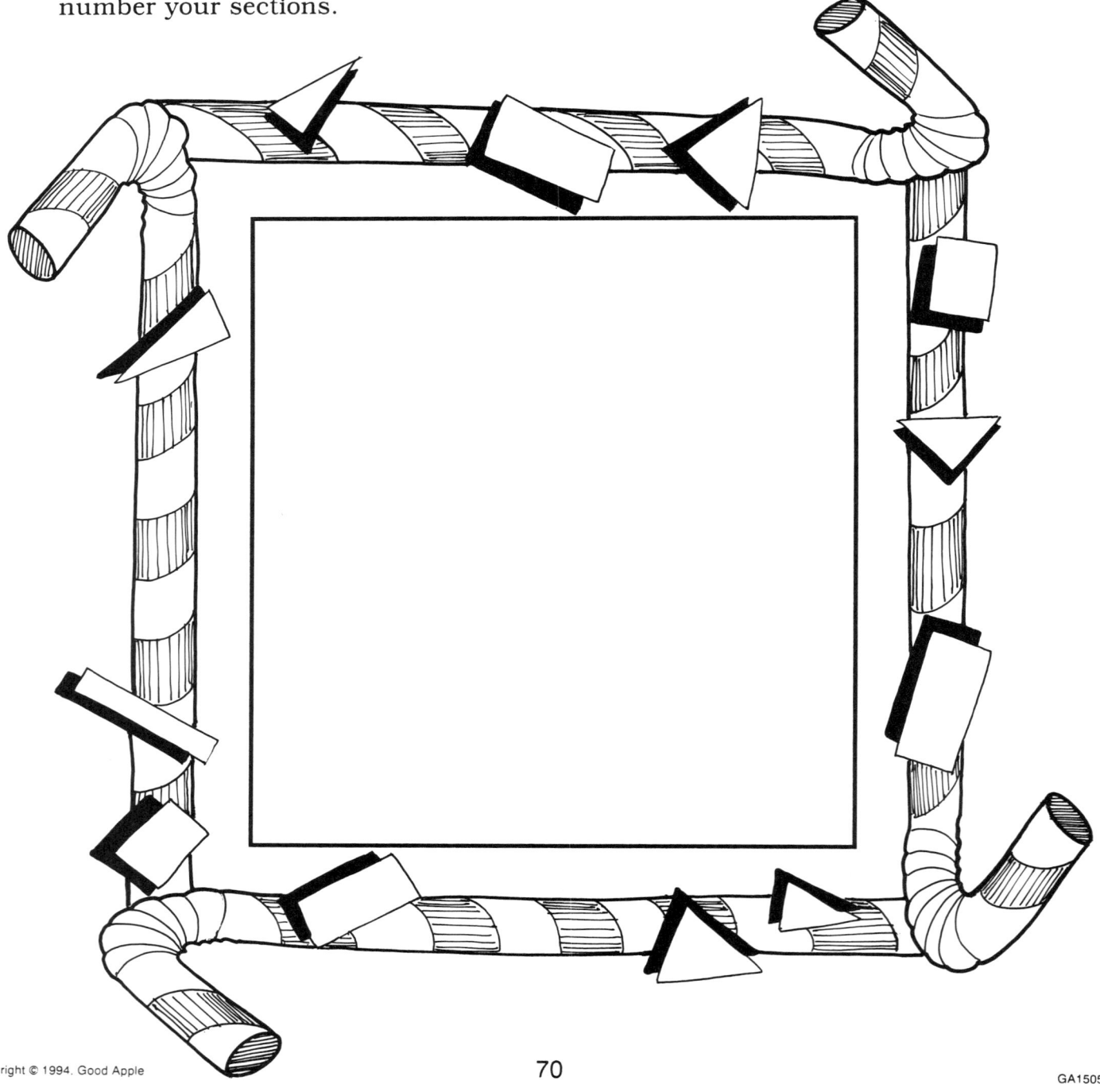

Visualizing the Construction of Cubes

# Box It Up!

Circle the shapes below that can be folded into closed boxes or cubes.

For the shapes you've circled, suppose that each one is folded into a cube with the X side on the bottom. Now put a T in the square that would be on top.

Spatial Relations

# A Pointed Problem

How many different ways can the block be placed into the box? Think about all the different ways in which the arrow could be pointed. Answer: _____

Identifying Congruent Figures,
Using Transformations

# Trace Case

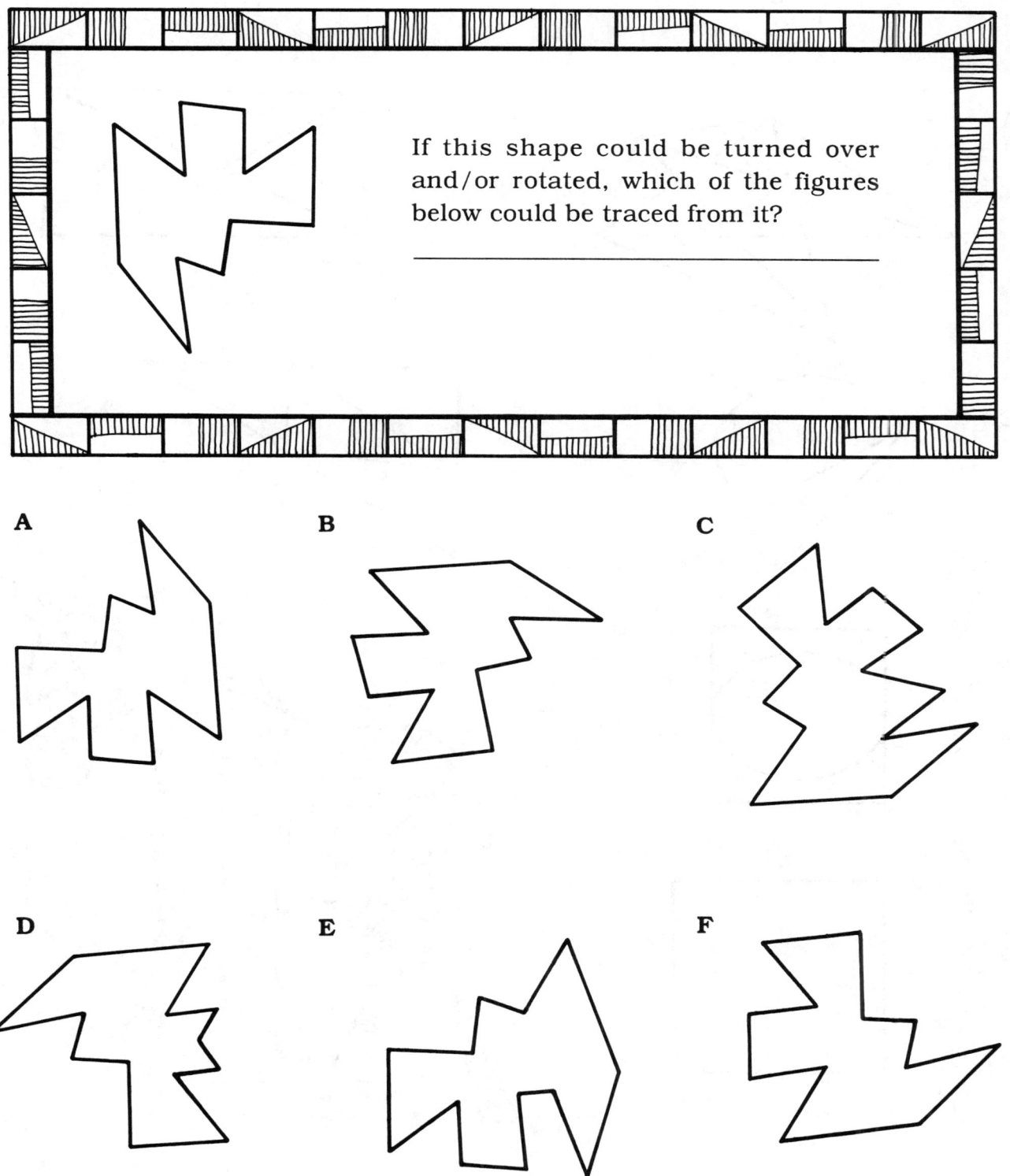

If this shape could be turned over and/or rotated, which of the figures below could be traced from it?

_____

A  B  C

D  E  F

Tracing Networks

# Network News

In which of these networks can you trace over all the lines without lifting your pencil or retracing your steps? Answers: Networks _____

A.

B.

C.

D.

E.

F.

Analyzing and Designing Networks

# More Network News

An eighteenth century mathematician, Leonhard Euhler, studied such networks. He proved that a network can be traced only if it has exactly two odd vertices or if it has no odd vertices.

What are these odd vertices? A vertex is a point where two or more lines intersect. A vertex is odd if an odd number of paths meet there. These are odd vertices:

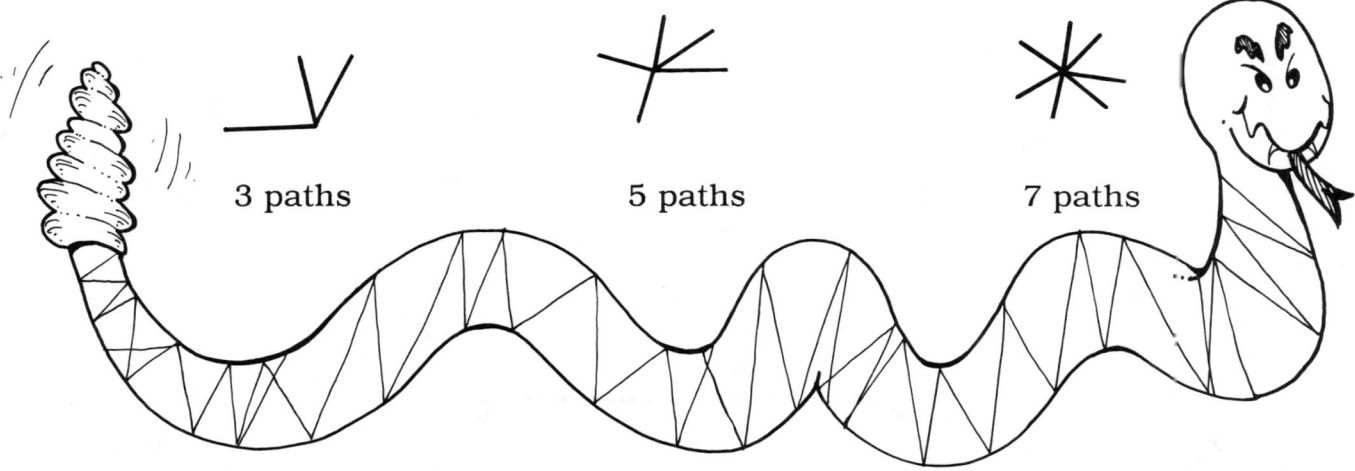

Look back at the networks on the previous page and count the odd vertices. Do your solutions agree with Euhler's findings? Use Euhler's rule to study this network.

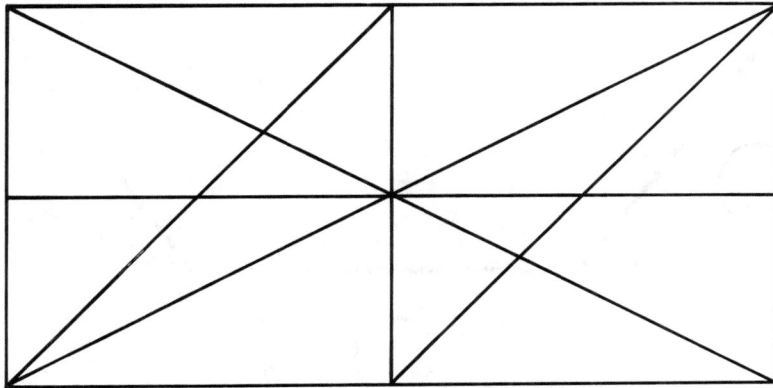

A. Can it be traced? _____

B. Why or why not? _____

**Now draw some networks of your own. Using Euhler's rule, draw some that can be traced and some that cannot be traced. Have a classmate check your work.

Topology, Interior and Exterior Spaces

# Spying on the Spies

Below is a sketch of the Hexagon, military headquarters for the citizens of Flambia. The circles outside the Hexagon represent six outposts built by the Zilches who are trying to spy on the Flambians. The squares represent spy outposts built by the Flambians to spy on the Zilch spies. Do you have that straight? Now the Flambians wish to build a high fence that would allow their outposts (the squares) access to the Hexagon but would bar access from the Zilch outposts (the circles) to the Hexagon. Can you draw such a fence?

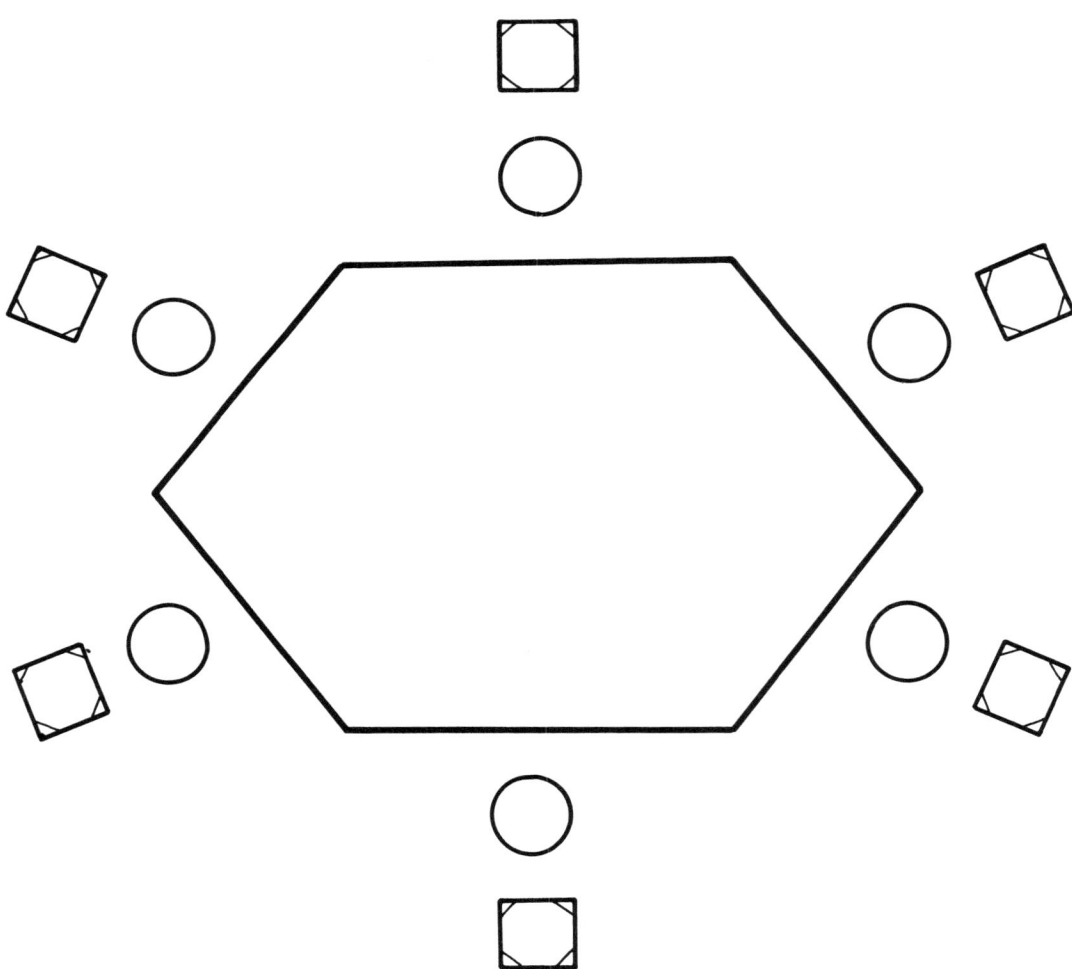

Visualization,
Constructing Squares

# Square Fare

Carefully cut out each shape. For each one, make one cut so that the two resulting pieces can be put together to form a square.

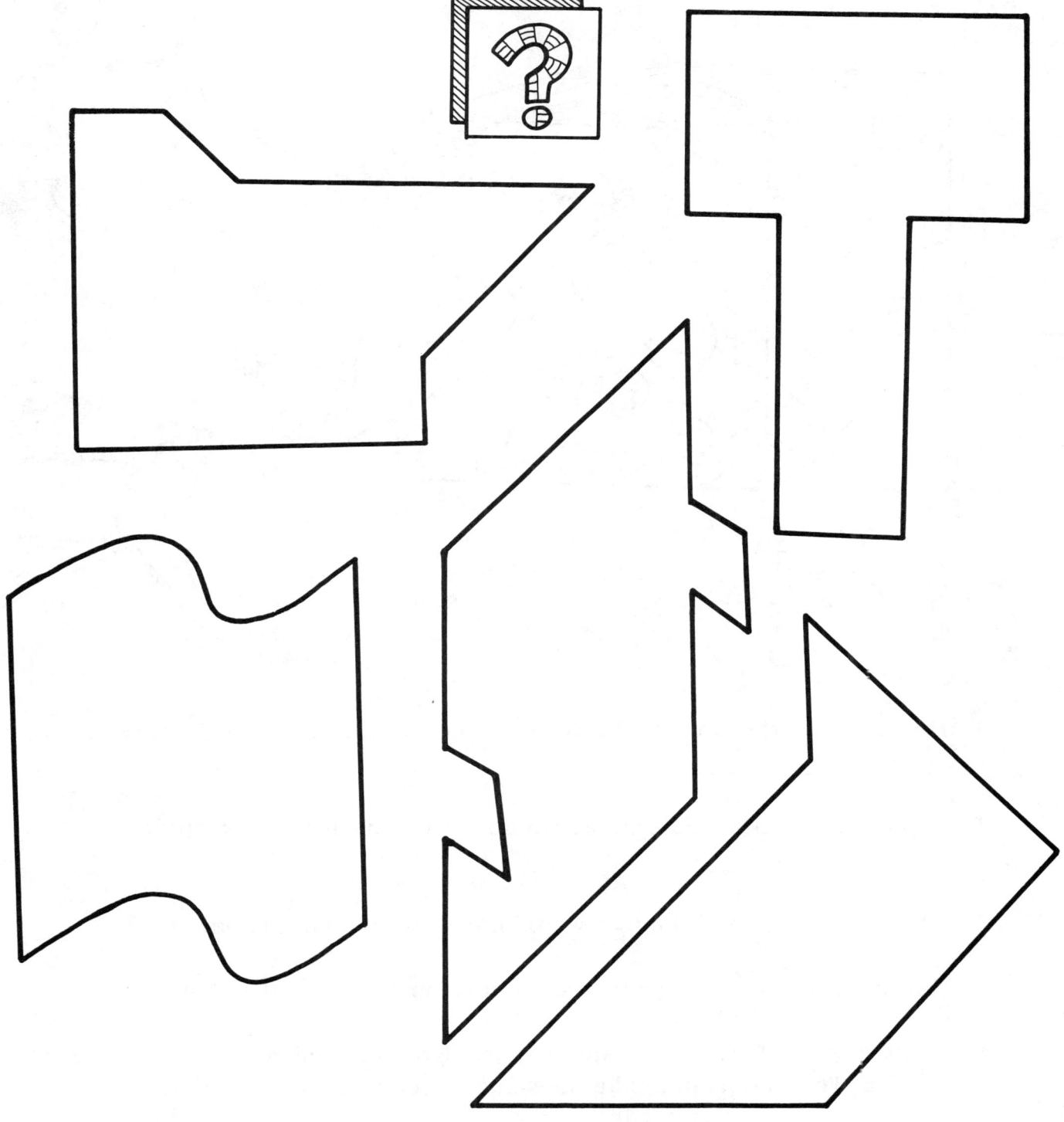

Finding and Analyzing Volume,
Exploring Properties of a Cube

## Cubic Questions

A. If this cube were sliced into small cubes, each 1 cm x 1 cm x 1 cm, how many small cubes would there be? _____

B. How many of the smaller cubes would have a design on exactly three sides? _____

C. How many of the small cubes would have a design on exactly two sides? _____

D. How many of the small cubes would have a design on just one side? _____

E. How many of the small cubes would have no design on them? _____

**A quick way to check if your answers are right is to add your answers in B, C, D, and E. Their sum should be the same as your answer in A. Why?

Constructing Triangles,
Recognizing Similar Triangles

# Triangle Trouble

Mrs. Yardley has placed nine stakes in her yard, represented here by nine dots. She has just asked her gardener to use a piece of string to rope off a triangle for a new flower bed. Each point of the triangle must be one of these stakes. How many *different* triangles can the gardener make? _____

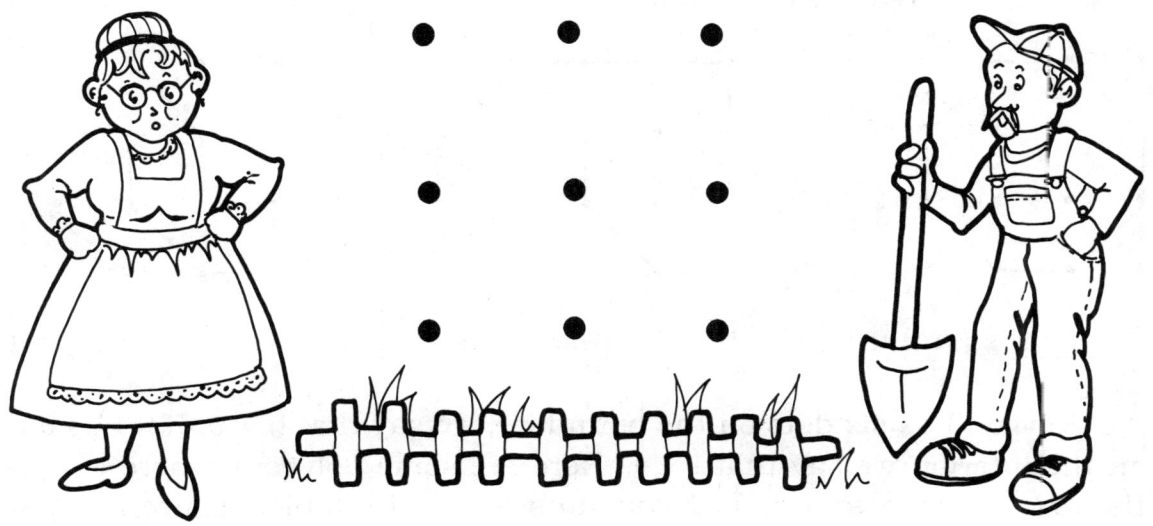

Use this space to draw more 3 x 3 dot squares for your different triangles. Remember two triangles are the same shape if one can be made to match the other by sliding, turning, or flipping.

Finding Area,
Devising Formulas

# Area Alert

Imagine that these shapes have been made on a geoboard.

1. If ▢ is one square unit, how many square units is △? _____

2. Now find the area of these shapes.

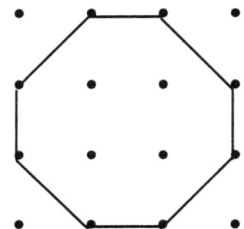 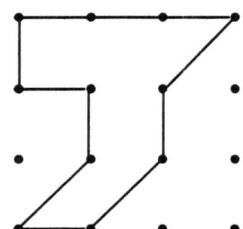 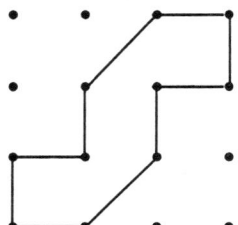

A. _____ sq. units    B. _____ sq. units    C. _____ sq. units

3. If a polygon has six dots on the boundary, we say that b = 6. If it has one dot on the interior, we say that i = 1. Here are some polygons where i = 0. Find the area of each and record your answer in the table. Figure A has been done for you.

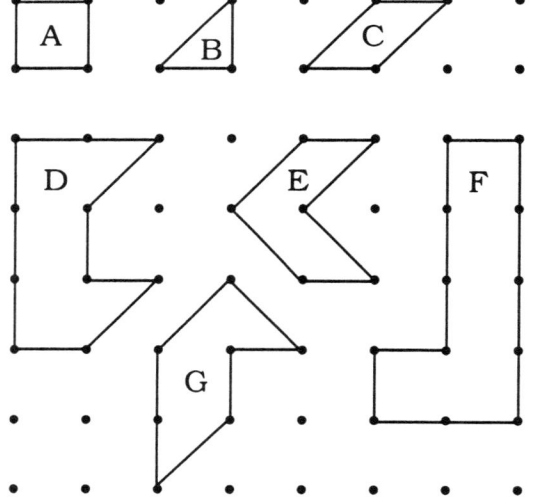

|    | Interior Dots | Boundary Dots | Area |
|----|---------------|---------------|------|
| A. | 0             | 4             | 1    |
| B. |               |               |      |
| C. |               |               |      |
| D. |               |               |      |
| E. |               |               |      |
| F. |               |               |      |
| G. |               |               |      |

4. **Can you write a formula that shows how to find the area of a polygon when the number of boundary dots is known and the number of interior dots is 0? _____

5. Here are some polygons with one interior dot. Find the area of each and record your answers in the table.

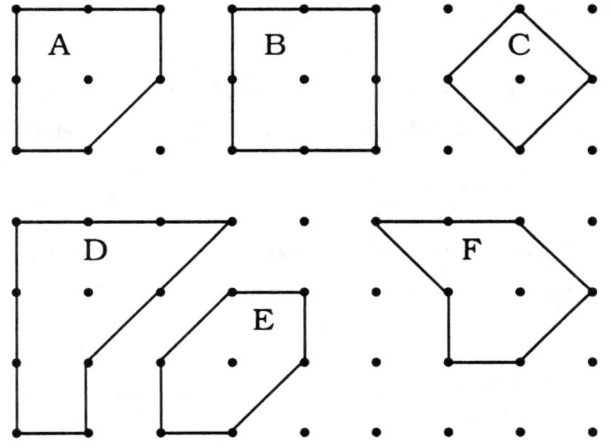

| | Interior Dots (i) | Boundary Dots (b) | Area (A) |
|---|---|---|---|
| A. | _____ | _____ | _____ |
| B. | _____ | _____ | _____ |
| C. | _____ | _____ | _____ |
| D. | _____ | _____ | _____ |
| E. | _____ | _____ | _____ |
| F. | _____ | _____ | _____ |

6. **Can you write a formula that shows how to find the area when the number of boundary dots is known and the number of interior dots is 1? _____

The formulas you have just discovered can be written as a single formula which relates the area (A) of *any* polygon to the number of dots inside the polygon (i) and the number of dots on its boundary (b). It is called Pick's rule and is written: $A = i + \frac{1}{2}b - 1$

7. Use Pick's rule to find the area of these polygons.

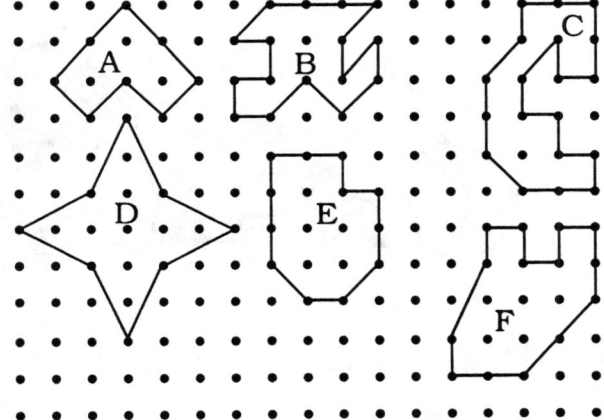

A. _____
B. _____
C. _____
D. _____
E. _____
F. _____

8. **Use Pick's rule to help you draw polygons with 3, 4, 5, 6, 7, 8, 9, and 10 sides so that each shape has the smallest possible area. Use dot paper or a geoboard.

Finding Perimeter,
Converting Feet to Yards

# Fencing Fido and Friends

The owners of three dogs–Fido, Sparky, and Rover–are putting up new fences around their backyards. From looking at the diagrams below, whose yard do you think needs the most fencing? _____

A. Now find the exact number of feet of fencing needed to enclose the backyards. Write your answers on the blanks.

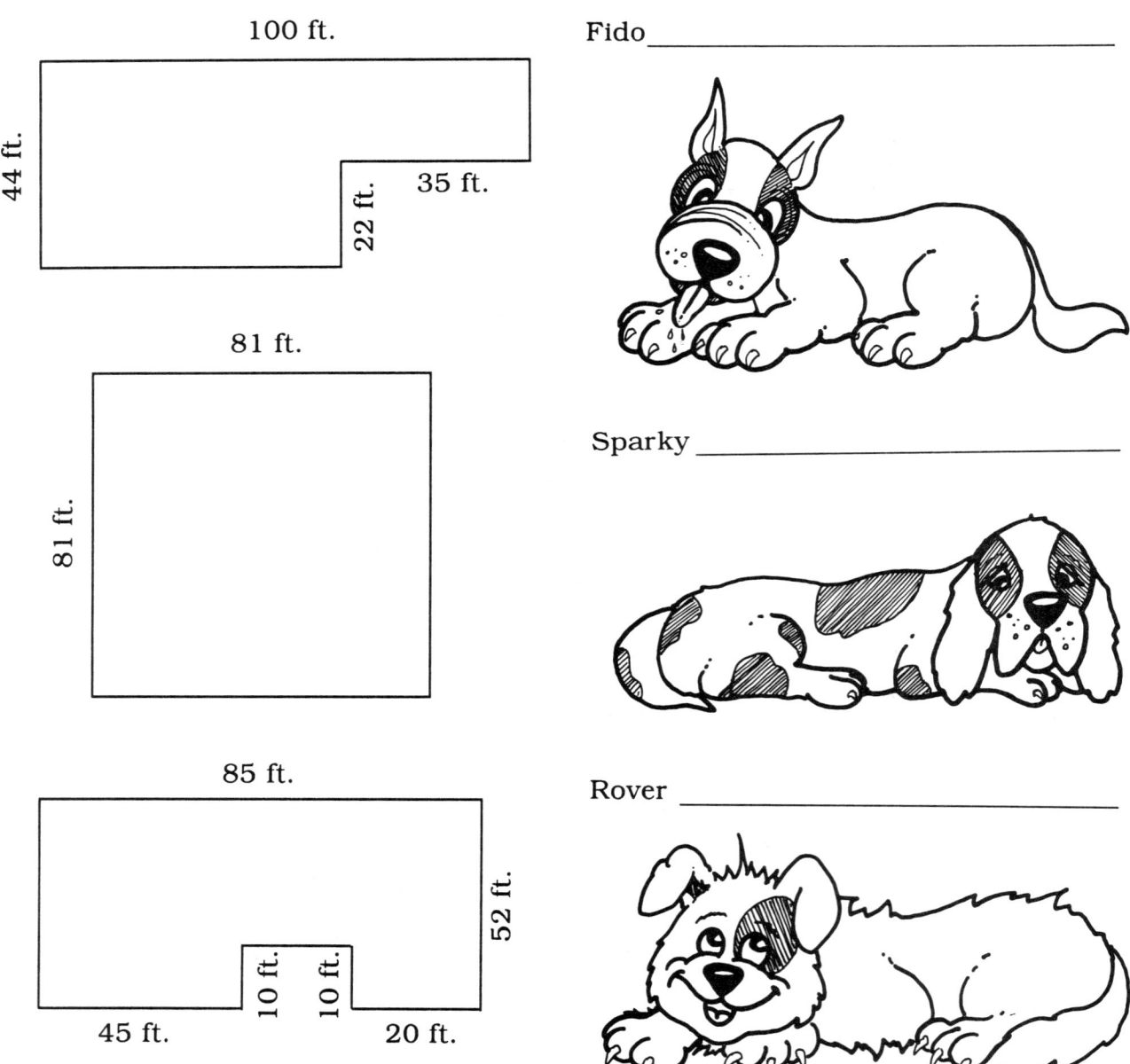

Fido _____

Sparky _____

Rover _____

B. If the fencing is sold by the *yard*, how many yards would be needed for each dog? Fido _____ Sparky _____ Rover _____

Finding Area, Anaylzing the
Relationship Between Perimeter
and Area of Various Shapes

# Fencing Finesse

Let's look at Sparky's fence again.

81 ft.

81 ft.

Sparky has just given birth to four puppies, and Sparky's owner, Bryce, wants to enlarge the yard. But Bryce has no more fencing to use. Bryce's dad suggests he can enclose a larger area with the same amount of fencing. Can you figure out what Bryce should do?

A. First find the area of the existing yard. _____

B. Now experiment by drawing different shapes with the same perimeter as Sparky's existing yard. Calculate the area of each until you find one with a larger area. Draw your answer here.

C. The area of Sparky's new yard is _____.

**D. By what percentage did the yard increase in size? _____

Converting Among English Units of Length, Area, and Volume

# How Far Is a Furlong?

Here is a quiz to test your knowledge of such English units of measure as furlongs, rods, leagues, acres, and pecks. First you'll need to do a little research to find the correct equivalences for any units you don't know. Then look at the three items in each row. Circle the one that is *longest* or *largest*.

1. A) 5 yards              B) 1 rod             C) 17 feet
2. 40 rods                 B) 670 feet          C) 200 yards
3. A) 9 furlongs           B) 1750 yards        C) 1 mile
4. A) 14,202 feet          B) 3 miles           C) 5,284 yards
5. A) 1 square yard        B) 1,308 sq. in.     C) 8 square feet
6. A) 1 sq. mile           B) 620 acres         C) 150 sq. rods
7. A) 1 acre               B) 43,000 sq. ft.    C) 5000 sq. yds.
8. A) 28 cubic ft.         B) 1 cubic yd.       C) 45,000 cubic in.
9. A) 1 peck               B) 8 quarts          C) 18 pints
10. A) 1 bushel            B) 5 pecks           C) 32 quarts

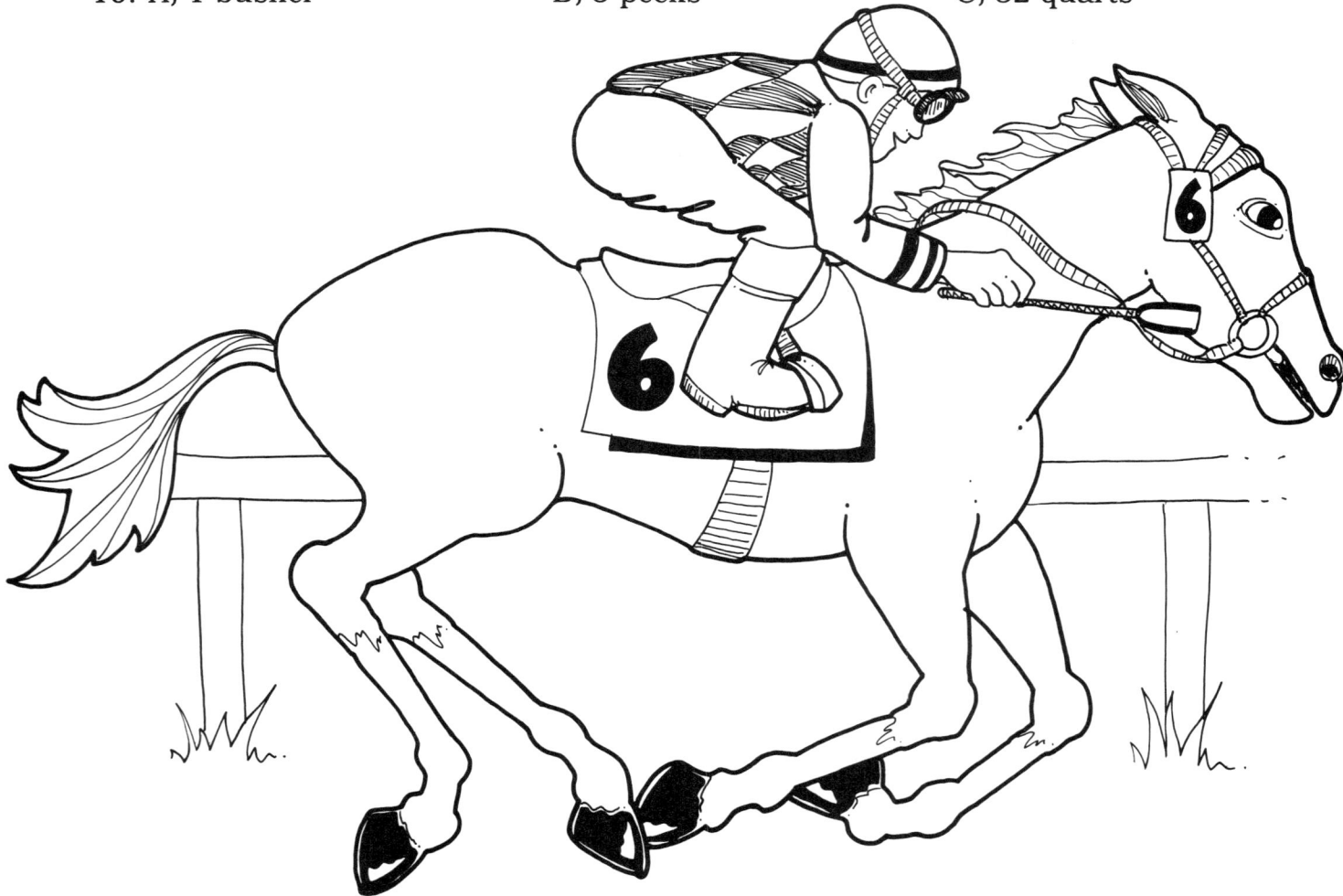

Measuring Height,
Estimating Products with Money

# Payday

You have just surprised your parents by cleaning the garage and your bedroom without being asked. Your dad is so thrilled that he offers to pay you. He gives you three choices of payment:

1. Your height in *quarters* laid end to end.

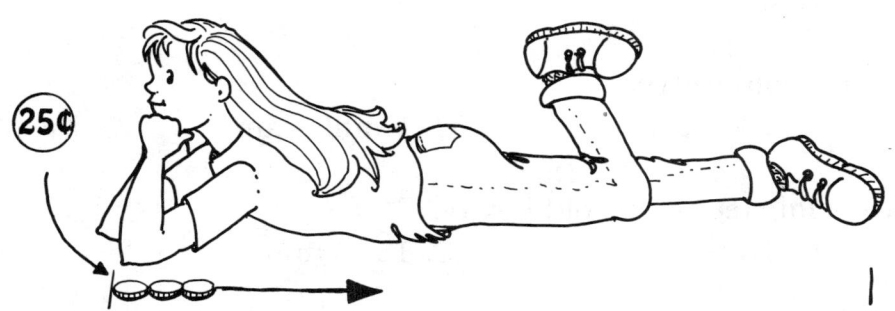

2. Your height in *dimes* laid end to end.

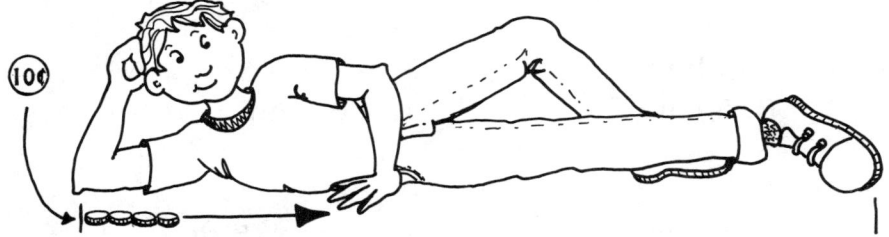

3. Your height in *nickels* stacked on top of one another.

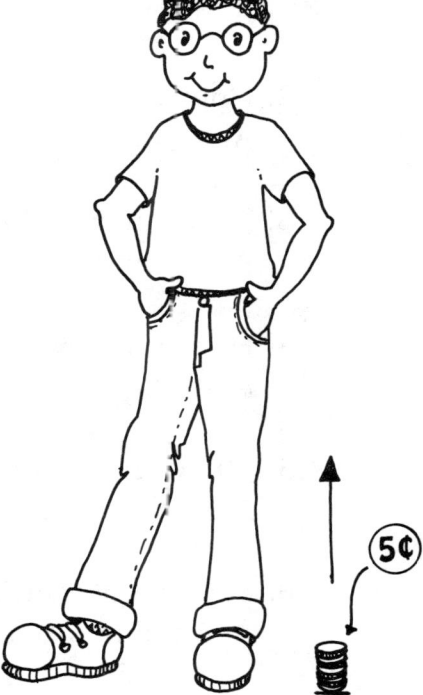

A. Which way do you think you'd get paid the most? _____

B. Measure your height in inches: _____

C. Measure: Number of quarters per inch (use diameter) _____

        Number of dimes per inch (use diameter) _____

        Number of nickels stacked per inch _____

D. How much money would you be paid in quarters? _____

    Dimes? _____ Nickels? _____

E. Were you right in Step A? _____

Estimating Length and Weight
in Metric Units

# Trivial Lengths and Widths

Here is a trivia/math quiz that requires you to know a little bit of trivia and a lot of math. For each question, circle the most reasonable answer–even if you have to do some guessing. Then try to research the facts necessary to verify the correct answers.

1. How many grams would a regular-sized box of graham crackers weigh?
   A. 4.54 g         B. 45.4 g         C. 454 g

2. How much would one paper clip weigh?
   A. 1 gram         B. 1 kilogram     C. 1 milligram

3. How tall would an average fourteen-year-old boy be?
   A. 15 meters      B. 157 cm         C. 1570 mm

4. How much would an average fourteen-year-old boy weigh?
   A. 48 kg          B. 4800 mg        C. 4800 g

5. The width of a tennis court is about
   A. 8 m            B. 8 cm           C. 800 mm

6. Most credit cards are about
   A. 8.6 mm x 5.4 mm   B. 8.6 cm x 5.4 cm   C. .8 m x .5 m

7. The length of the Rio Grande is about
   A. 3040 m         B. 304,000        C. 3040 km

8. The height of Mt. Everest is about
   A. 870 m          B. 870 km         C. 870 cm

9. An elephant could weigh about
   A. 45 kg          B. 4500 g         C. 4500 kg

10. A cat could weigh about
    A. 3 kg          B. 30 kg          C. 30 g

11. What is the diameter of the earth?
    A. 12.756 km     B. 12,756 km      C. 1,275,600 km

12. The deepest lake in the U.S. is Crater Lake, Oregon. About how deep is it?
    A. 589 m
    B. 5,890 m
    C. 58,900 mm

13. The longest bridge span in the U.S. is the Verrazano-Narrows in New York. Its length is about
    A. 12.78 m
    B. 1278 m
    C. 127,800 m

14. The highest U.S. bridge is the Royal Gorge in Colorado. How high above water is it?
    A. 321 cm
    B. 3.21 m
    C. 321 m

15. The longest river in the world is the Nile. Its length is about
    A. 667.1 km
    B. 6671 km
    C. 66.71 km

16. The highest mountain in Africa is Kilimanjaro. Its tallest point is about
    A. 5895 m
    B. 58.95 m
    C. 589.5 m

Estimating and Measuring
Length, Weight, and Volume in
Unfamiliar Units

# Long Gone

Listed here are a few units of measure left over from days gone by. Many were used in ancient times.

3 inches = 1 palm
4 inches = 1 hand
6 inches = 1 span
18 inches = 1 cubit

1 omer = .45 peck
= 3.964 liters
ephah = 10 omers
shekel = .497 ounce
= 14.1 g

1. Find your height in palms.

    Estimate _____ Measure _____

2. What would your height be in *spans*? _____

3. Find the width of your classroom.

    Estimate number of hands _____ Measure number of hands _____

    Estimate number of cubits _____ Measure number of cubits _____

4. About how many omers would it take to fill a bushel? _____

5. About how many shekels would it take to make 1 pound? _____

6. Which is more, 2 ephahs or 75 liters? _____

7. Which is more, 1 kilogram or 70 shekels? _____

Computing and Comparing
Volume of Solids

# Billy's Chili

Billy is a salesman who's trying to promote his latest product–hot, spicy chili. He wants to package it in a container that holds as much as possible. Which container below will hold the *most* chili?

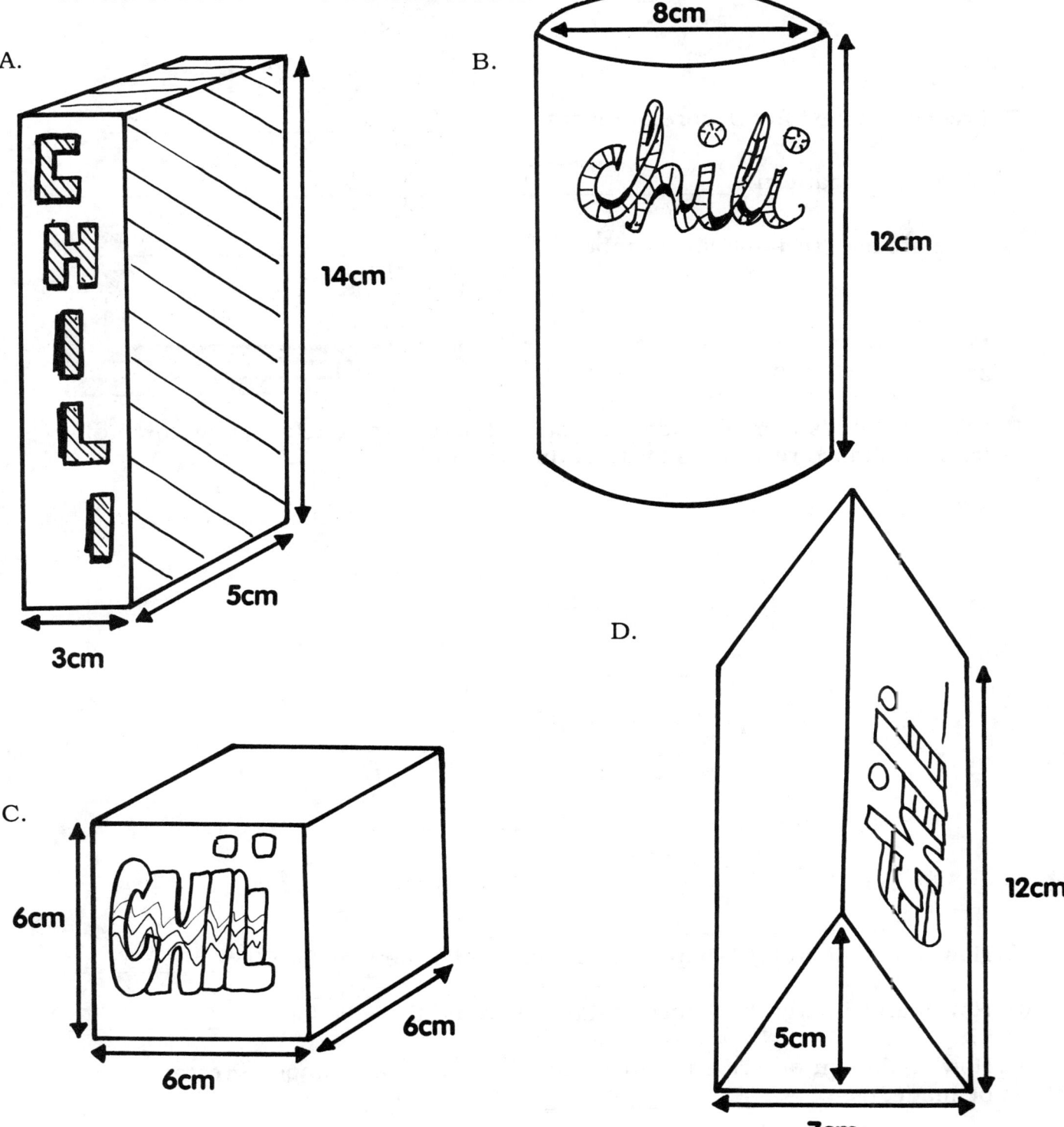

Recognizing Number Patterns
Working with Multiples

# Shapely Numbers I

A. Here are some "square" numbers: Draw the next three square numbers.

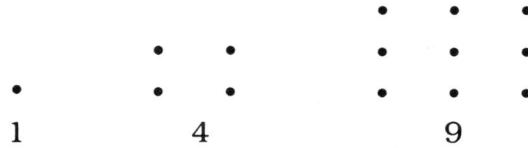

1      4      9

B. Predict the next five square numbers _____ _____ _____ _____ _____

C. What is the pattern?

D. Here are few "rectangular" numbers.

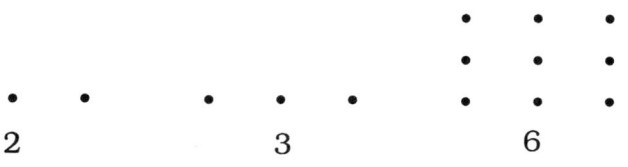      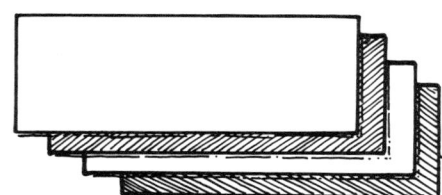

2      3      6

E. Some numbers can be made into many different rectangles. Draw three differently shaped rectangles for the number 24.

F. How many differently shaped rectangles would there be for 60? _____

G. How many differently shaped rectangles would there be for 72? _____

H. How could you describe numbers for which only one shape of rectangle can be made? _____

Recognizing Number Patterns

# Shapely Numbers II

I. Here are some *triangular* numbers.

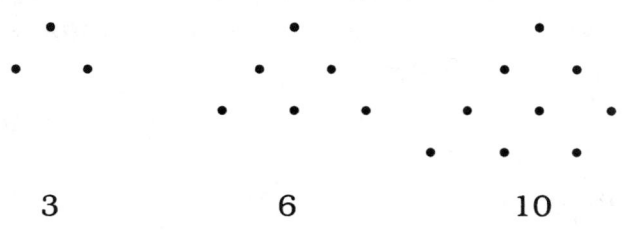

   3           6          10

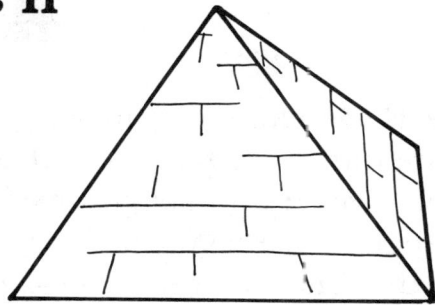

Draw the next three triangular numbers.

J. Predict what the next three triangular numbers will be. _____  _____  _____

K. What is the pattern? _____

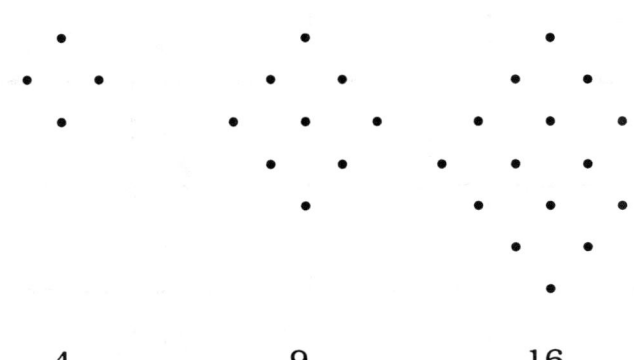

   4           9          16

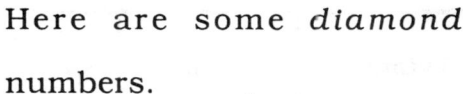

Here are some *diamond* numbers.

L. Draw the next two diamond numbers.

M. Predict the next two diamond numbers. _____  _____

N. What is the pattern? _____

Problem Solving

# Campground Confusion

It is Monday morning at the Birdseye View Campground, and the counselors are frantically trying to schedule the week's athletic activities. The campers will be divided into six groups: the Orioles, Robins, Eagles, Wrens, Cardinals, and Sparrows. Each group needs to participate in each activity once during the week. The counselors have also received these requests.

The Robins should swim on Monday.

The Cardinals can't canoe on Wednesday or Thursday.

The Orioles would like to be the first ones on the softball diamond and the last ones on the hiking trail.

Can you help the counselors complete this schedule?

|  | Monday | Tuesday | Wednesday | Thursday | Friday | Saturday |
|---|---|---|---|---|---|---|
| Archery |  |  |  |  |  |  |
| Canoeing |  |  |  |  |  |  |
| Horseback Riding |  |  |  |  |  |  |
| Swimming | Robins |  |  |  |  |  |
| Hiking |  |  |  |  |  |  |
| Softball |  |  |  |  |  |  |

Problem Solving,
Geometry

# Picky, Picky

Twenty-four toothpicks have been laid on a table in this arrangement:

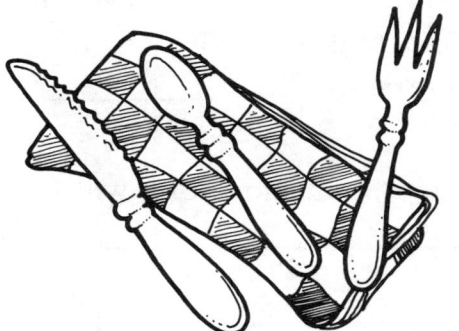 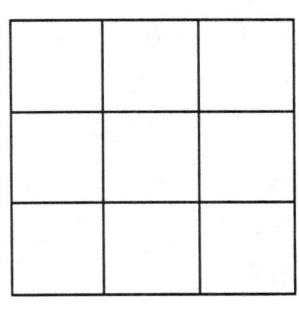 

1. How many squares in all have been formed by these toothpicks? _____

2. Try to take out four toothpicks in a way that will leave nine squares. Draw your answer here.

3. Replace all twenty-four toothpicks in their original places. Now take out four toothpicks in a way that will leave seven squares. Draw your answer here.

4. Use eight toothpicks to make a shape containing two squares and four triangles.

5. Use seven toothpicks to make a shape containing two squares and three triangles.

Number Play,
Logic

# Tiny Teasers

1. Mrs. Fashion wanted to order a new wardrobe from her favorite mail order catalog. Unfortunately, she discovered her dog, Button, had torn several pages from her catalog. Mrs. Fashion could not find pages 35, 36, 54, 55, 103, and 104. How many pieces of paper did Button remove? _____

2. If it takes 6 people 4 days to dig a ditch that's 10 feet long, how long will it take 3 people to dig a ditch 5 feet long? _____

3. A duck family is crossing the street. There are 2 ducks in front of a duck, 2 ducks behind a duck, and a duck in between 2 ducks. How many ducks are there altogether? _____

4. How many times in 24 hours will a digital clock read the same forwards and backwards? Use this space to write out the possibilities.

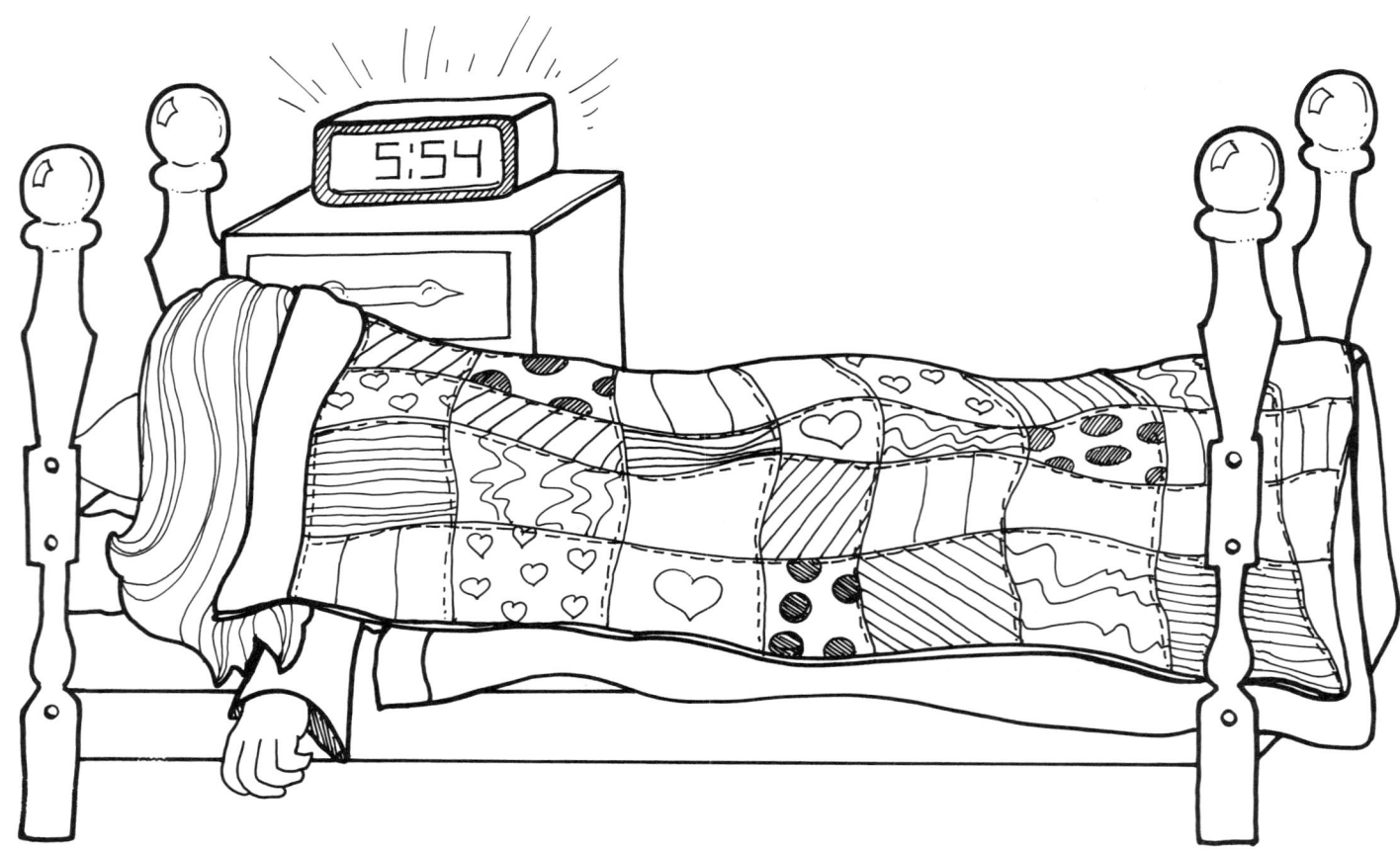

5. An eight-digit number contains two 1s, two 2s, two 3s, and two 4s. The 1s are separated by one digit, the 2s by two digits, the 3s by three digits, and the 4s by four digits. Can you write this number? _____

6. Draw the face of a watch. Put two straight lines across the face (dividing it into three sections) so that the numbers in each section add up to the same sum.

7. If it takes 50 minute for 1 wet pair of pants to dry on a clothesline, how long will it take 5 pairs of pants to dry? _____

8. Three sisters have nine chocolates weighing 6, 12, 4, 9, 8, 15, 5, 13, and 10 ounces each. Shari's 3 chocolates weigh twice as much as Mary's 3 chocolates. Teri's three chocolates weigh more than either Mary's or Shari's. Which sister has which chocolates?

Shari _____
Mary _____
Teri _____

Logic, Scheduling

# Publisher's Paradise

Audrey Author is spending the day at the national headquarters of Publisher's Paradise, preparing to publish her first book. She needs to spend two hours with her editor, two hours with her illustrator, one hour with the book designer, and one hour with the marketing director. Audrey would also like to spend an hour with her best friend who is a secretary to the publisher. All of Audrey's appointments will be held in the same office building, which is open from 9 a.m. until 5 p.m. Can you help Audrey plan her schedule for the day, using the following information?

1. Audrey's illustrator is just returning from an art exhibit and won't be available until 1:00 p.m.
2. Audrey knows she will need a one-hour break before her last meeting to take notes and rest.
3. All the secretaries are kept extremely busy except during their one-hour lunch break which begins at noon.
4. The designer and marketing director both leave at 2 p.m.
5. The editor has a private meeting with the designer from 11 a.m. to 12 noon.

### Schedule

9:00 _____
10:00 _____
11:00 _____
12:00 _____
1:00 _____
2:00 _____
3:00 _____
4:00 _____
5:00 Building closes.

Logic, Addition and Subtraction

# Puzzling Pizzeria

Here's the menu from Pete's Pizzeria.

|  | 12" | 14" | 16" | Extra Items | |
|---|---|---|---|---|---|
| Cheese pizza | $6.00 | $7.00 | $8.00 | Ham | Onion |
| Cheese and one item | 6.50 | 7.50 | 8.50 | Sausage | Gr. Peppers |
| Cheese and two items | 6.80 | 7.80 | 8.80 | Pepperoni | Bl. Olives |
| Cheese and three items | 7.10 | 8.10 | 9.10 | Grnd. Beef | Mushrooms |
| Deluxe (8 toppings) | 7.50 | 8.50 | 9.50 | | |

Can you tell what I ordered from Pete's Pizzeria by using these clues?

1. I ordered two pizzas.
2. I always choke on green peppers, and I'm allergic to olives.
3. One pizza had one extra item.
4. Both pizzas had pepperoni on them.
5. One pizza cost $1.30 more than the other.
6. Neither pizza contained more than one meat.
7. Pete was out of onions.
8. I received less than $5.00 back from my $20.00 bill.

Logic, Computing Volume

# Zams for Sale I

Aliens from Pluto have landed in your backyard, and they have set up a factory that produces Zams. The aliens are now packaging the Zams for wholesale.

Use this information to help the aliens figure out how many Zams will fit inside each xob (box).
1. One kray equals three geef.
2. Each xob is 3 geef high and 2 geef wide. It is 1 kray deep.
3. Each Zam is 1 geef high, 1 geef wide, and 1 geef deep.

Sketch the xob here and label the sizes of each dimension.

How many Zams will fit inside each xob? _____

Logic, Simple Computations

# Zams for Sale II

Sophie decided to purchase some Zams from the aliens, but their money system is a little strange. They wished to be paid with zucchini and fruitcakes.

The aliens told your friend Sophie that:
    A. One fruitcake is worth 3 zucchini.
    B. Each Zam costs 4 zucchini.

Sophie has 5 fruitcakes and 5 zucchini. How many zams can she buy? _____

Logic

# Rope Riddle

Your mom has asked you to mark a garden plot in the backyard that is 8 yards square. You can't find a yardstick or a tape measure, but you do have a long piece of rope and a piece of chalk to mark it with. You know that the storage shed in your backyard is 7 yards long and 5 yards wide. How can you carry out your mother's request with this equipment and information?

_____
_____
_____
_____
_____
_____
_____

Logic

# Christmas Candy Confusion

It's Christmas Eve and you've just bought your six favorite people large boxes of candy. Each one is filled with a different assortment of chocolates, mints, rock candy, etc. As you're wrapping the gifts, you get a phone call from the candy shop owners. "One of those boxes you just purchased was filled with substandard candy," you are told. "Each piece of candy in each box is supposed to weigh 10 grams, but somehow it seems that one of your boxes was filled with candies weighing only 9 grams." The store owner asks you to bring back the faulty box for an immediate replacement.

The store closes in fifteen minutes, and tomorrow is Christmas, so you have to hurry. How can you find the light box in only one weighing?

_____
_____
_____
_____
_____
_____

# Answer Key

**Ten to Eighteen**, Page 1
Here is one solution. Variations are possible.

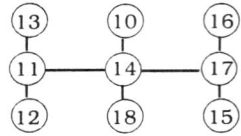

**Magic Formulas**, Page 2
Formula A: The answer is always 2.
Formula B: Your answer is always your original number.
Formula C: The answer is always 9.

**Sixty Sense**, Page 4
Here is one set of combinations that uses all numbers. There may be others as well.

| | | |
|---|---|---|
| 25 | 18 | 17 |
| 28 | 14 | 18 |
| 12 | 24 | 24 |
| 19 | 13 | 28 |
| 20 | 20 | 20 |
| 36 | 16 | 8 |
| 30 | 15 | 15 |
| 22 | 17 | 21 |
| 26 | 11 | 23 |
| 17 | 12 | 31 |
| 11 | 30 | 19 |
| 23 | 27 | 10 |

**Number Pals**, Pages 5-9
Part One:
1. one-step
2. three-step
3. two-step
4. one-step
5. three-step
6. two-step
7. three-step
8. seven-step
9. four-step
10. one-step

**All one-step numbers have two things in common:
a. the middle digit is less than 5 and
b. the sum of the first and third digits is less than 10
Other one-step numbers include 410, 346, 207, etc.
***This number turns into the palindrome 5,158,778,515 after seventeen steps!

Part Two:
a. One-step numbers are those where the first and third digits have a difference of 1.
b. Two-step numbers are those where the first and third digits have differences of either 5 or 6.
c. 1, 2, and 3 are three-step numbers.
   4, 5, and 6 are four-step numbers.
   7, 8, and 9 are five-step numbers.
d. In three-step numbers, the first and third digits have differences of 3 or 8.
e. In four-step numbers, the first and third digits have differences of 4 or 7.
f. In five-step numbers, the first and third digits have differences of 2 or 9.

**Sign In**, Page 10
Other answers may be possible.
1. 15 ÷ (2 + 3) = 3
2. (0 x 7) + 6 = 6
3. [(3 x 9) + 1] ÷ 4 = 7
4. [(5 x 9) + 15] x 3 = 9
5. 4 + [2 x (4 x 2)] = 20
6. 4 - [2 ÷ (4 ÷ 2)] = 3
7. (3 + 4 + 4) x 2 = 22
8. [(7 + 6) x 2] + 4 = 30
9. [(20 + 15) ÷ 5] - 2 = 5
10. 20 - [15 ÷ (5 - 2)] = 15
11. [[1 + (2 x 3)] + 4] - 5 = 6
12. 10 x [9 ÷ [(8 + 7) - 6]] = 10

**What's the Problem?** Page 11
Other answers are possible.
1.  865      3.   78      5.   159
   + 37          x  9          x   3

2.  123      4.  906      6.  3845
   -  45          x  8          x   7

**Daring Dots I**, Page 12
a.   35    b.   66
   x 60       x 22
   2100      1452

**Darling Dots II**, Page 13
c.    41    d.   5465
   62)2542    x    22
              120,230

Since these problems can be so difficult, the teacher may wish to give students a clue by revealing at least one digit of the solution.

**Cookie Contest**, Page 14
The diagram shows which cookies would be covered by the top one.
Cookies that do touch: 5 x 11 x 3 x 7 x 2 x 4 x 10 x 2 = 184,800
Cookies that don't touch: 13 x 6 x 9 x 12 x 6 x 4 = 202,176
Cookies that don't touch have the larger product.

**Number Clues**, Page 15
1. 19, 2. 63, 3. 55, 4. 61, 5. 95 (5 x 19)

**Disappearing Digits**, Page 16

| 1. 295 | 2. 201 | 3. 19 | 4. 03 |
|---|---|---|---|
| 968 | 5 | 789 | 8482 |
| 6 | 15 | 37 | 338 |
| 8 | 44 | 9 | 252 |
| | 16 | 74 | |
| | | 7 | |
| | | 8 | |
| | | 9 | |
| | | 259 | |

**Logical Letters**, Page 17
A = 3     B = 6     C = 9
D = 1     E = 5     F = 4
G = 7     H = 8     K = 2

**A"maze"ing Mathematics I**, Page 18

```
↑
7 ——— 14 ——— 21      25
9       35    28      30
52      42    40      36
55      49    60      48
65      56 ——— 63 ——— 70
```

The path can be drawn by connecting multiples of 7.

**False Advertising**, Page 19
Bonus words: minus, sum, add

How of(ten) have you wanted to use an umbrella but didn't because it was too bulky and inconvenient? What you need is now available! You've seen the(m in us)e at Niagara Falls, in rain forests, and ye(s, even) in hurricanes. And now what you need is being produced in the U.S.A.! It's the Amazing Collapsible Umbrella produced by the Sta-Dri Company in Es(six)New Mexico. Thi(s um)brella has been tested in the rai(n in e)very state in the nation. Because of new technology using special pulleys wi(th ree)ls of plastic fibers, each o(f our) umbrellas stretches to a h(eight)of 4 feet and shrinks to a size that will slip into your back pocket. I(t wo)n't be long before you'll see our umbrellas (on e)very block. Get in on the latest f(ad. D)on't delay–order yours today! (Endorsed by world champion surfer, Whet Waver, who says, "I(f I've)ever found a product for keeping dry, this is it!")

**Spelling Text**, Page 20
a. 30
b. ninety
c. thousand
d. 18
e. twelve
f. 13
g. eleven
h. 4
i. 20
j. hundred
k. fourteen
l. 15
Total Score: 100

**Check It Out**, Page 21
1. *Fifth* misspelled
2. *February* misspelled
3. There are not 30 days in February.
4. A comma is needed between city and state.
5. *Michigan* misspelled
6. *Company* misspelled
7. *hundred* misspelled
8. *forty* misspelled
9. *eight* misspelled
10. *dollars* misspelled
11. Cents differ from dollar box to written line
12. Name in signature is different from name printed on check.

**Box Baffler**, Page 22

| -7 | -6 | 4 | 3 | -4 |
| 2 | -7 | -5 | 5 | 0 |
| 3 | -5 | 1 | -8 | 9 |
| -5 | 5 | 2 | 4 | -1 |
| -3 | 8 | -2 | 1 | 6 |

**More or Less**, Page 23
Start

| 1 | -3 | -6 | -2 | -1 | 1 | 3 |
| 3 | 0 | 4 | -3 | 0 | 4 | -2 |
| 5 | -4 | -8 | -6 | -4 | -1 | 2 |
| -1 | 0 | -5 | -2 | -1 | -4 | 5 |
| 1 | -3 | -6 | -9 | -5 | -7 | 1 |
| -3 | -7 | -2 | 0 | 2 | -3 | 4 |
| -13 | -11 | -8 | -5 | -2 | 1 | 7 |
| -10 | -7 | -4 | 0 | 1 | 6 | 10 |

Finish

**Boswell's Beanbags**, Page 24
Other answers are sometimes possible.
-36 is not possible.

| -32 | = | -6 | -6 | -6 | -7 | -7 |
| -21 | = | -7 | -7 | -5 | -5 | -3 |
| -9 | = | -6 | -7 | -7 | 3 | 8 |
| -7 | = | -5 | -6 | -7 | 3 | 8 |
| -1 | = | -5 | -7 | -7 | 8 | 10 |
| 0 | = | -5 | -6 | -7 | 8 | 10 |
| 4 | = | -5 | -5 | -6 | 10 | 10 |
| 10 | = | -5 | -6 | 3 | 8 | 10 |
| 15 | = | -5 | -6 | 8 | 8 | 10 |
| 30 | = | 3 | 3 | 8 | 8 | 8 |

32 is not possible.

**Bingo Banger**, Page 25

| -97 | -97 | -92 | -92 | -96 |
| -98 | -96 | -98 | -96 | -96 |
| -93 | -97 | -96 | -102 | -100 |
| -90 | -96 | -98 | 96 | -94 |
| -96 | -96 | -95 | -102 | -98 |

**Riddle Rattler**, Page 26
Because it had SO MANY PROBLEMS.

**Prime Path**, Page 27
b. 234
c. 38

Start

| 2 | 46 | 8 | 31 | 41 | 9 | 10 | 24 |
| 3 | 5 | 11 | 13 | 17 | 13 | C 6 | 27 |
| 22 | 29 | 7 | 14 | 15 | 5 | 29 | 4 |
| 16 | 51 | 25 | 28 | 27 | 37 | 19 | 21 |
| 13 | 26 | 53 | 22 | 47 | 53 | 12 | 20 |
| 34 | 59 | 52 | 18 | 37 | 71 | 23 | 57 |
| 11 | 47 | 67 | 43 | 32 | 68 | 61 | 7 |
| 49 | 33 | 61 B | 63 | 57 | 39 | 53 | 59 |

Finish

**Prime Time**, Page 28
Other answers are also possible.
1. 7 + 13
2. 13 + 17
3. 17 + 23
4. 3 + 3
5. 17 + 19
6. 19 + 29
7. 17 + 37
8. 5 + 61
9. 31 + 47
10. 23 + 61
11. 23 + 67
12. 41 + 59
13. 41 + 83
14. 37 + 97
15. 59 + 89

It is always possible to write an even number as the sum of two primes.

**Number Fits**, Page 29
Multiples of 6: 18, 36, 42
Multiples of 7: 21, 28, 49
Odd numbers: 13, 27, 29
Numbers with sum of 40: 6, 14, 20

**Exponent Express**, Page 30

| | | | |
|---|---|---|---|
| $10^6 + 1$ | $10^7 - 10^6$ | $10^2 \times 10^2$ | $10^5$ |
| $10^4 + 10^5$ | $10^2 \times 10^2 \times 10^1$ | $10^3 \times 10^4$ | $10^3 + 10^3$ |
| $10^5 + 10^5$ | $10^6 - 10^0$ | $10^5 + 10^2$ | $10^7$ |
| $10^4$ | $10^3 \times 10^3 \times 10^1$ | $10^6 + 10^1$ | $10^3 \times 10^2$ |
| $10^7 - 10^4$ | $10^6$ | $10^7 - 10^7$ | $10^6 \times 10^0$ |
| $10^4 + 10^1$ | $10^7 \times 10^0$ | $10^8 - 10^7$ | $10^4 + 10^4$ |
| $10^3 \times 10^2 \times 10^0$ | $10^4 + 10^4$ | $10^6 - 10^1$ | $10^3 \times 10^2 \times 10^2$ |

Start → / Finish →

**Exponents for Experts I**, Page 31
1. 4, 1
2. 3, 7
3. 8, 5
4. 9, 2, 6

**Exponents for Experts II**, Page 32
1. 1
2. 5
3. 9
4. 4
5. 3
6. 2
7. 6
8. 7, 8

**Mystery Math**, Page 33
This system is based on the clock.
The number 12 acts like zero in base ten.

| + | 1 | 2 | 3 | 4 | 5 | 6 | 7 | 8 | 9 | 10 | 11 | 12 |
|---|---|---|---|---|---|---|---|---|---|---|---|---|
| 1 | 2 | 3 | 4 | 5 | 6 | 7 | 8 | 9 | 10 | 11 | 12 | 1 |
| 2 | 3 | 4 | 5 | 6 | 7 | 8 | 9 | 10 | 11 | 12 | 1 | 2 |
| 3 | 4 | 5 | 6 | 7 | 8 | 9 | 10 | 11 | 12 | 1 | 2 | 3 |
| 4 | 5 | 6 | 7 | 8 | 9 | 10 | 11 | 12 | 1 | 2 | 3 | 4 |
| 5 | 6 | 7 | 8 | 9 | 10 | 11 | 12 | 1 | 2 | 3 | 4 | 5 |
| 6 | 7 | 8 | 9 | 10 | 11 | 12 | 1 | 2 | 3 | 4 | 5 | 6 |
| 7 | 8 | 9 | 10 | 11 | 12 | 1 | 2 | 3 | 4 | 5 | 6 | 7 |
| 8 | 9 | 10 | 11 | 12 | 1 | 2 | 3 | 4 | 5 | 6 | 7 | 8 |
| 9 | 10 | 11 | 12 | 1 | 2 | 3 | 4 | 5 | 6 | 7 | 8 | 9 |
| 10 | 11 | 12 | 1 | 2 | 3 | 4 | 5 | 6 | 7 | 8 | 9 | 10 |
| 11 | 12 | 1 | 2 | 3 | 4 | 5 | 6 | 7 | 8 | 9 | 10 | 11 |
| 12 | 1 | 2 | 3 | 4 | 5 | 6 | 7 | 8 | 9 | 10 | 11 | 12 |

**Base Two Clues**, Page 34
C = 11
D = 100
E = 101
F = 110
G = 111
H = 1000
I = 1001
J = 1010
K = 1011
L = 1100
M = 1101
N = 1110
O = 1111
P = 10000
Q = 10001
R = 10010
S = 10011
T = 10100
U = 10101
V = 10110
W = 10111
X = 11000
Y = 11001

**A Weighty Situation**, Page 35
1. 1
2. 3-1
3. 3
4. 3 + 1
5. 9 - (1 + 3)
6. 9 - 3
7. (9 + 1) - 3
8. 9 - 1
9. 9
10. 9 + 1
11. (9 + 3) - 1
12. 9 + 3
13. 9 + 1 + 3
14. 27 - (1 + 3 + 9)
15. 27 - (3 + 9)
16. (27 + 1) - (3 + 9)
17. 27 - (1 + 9)
18. 27 - 9
19. (27 + 1) - 9
20. 27 - (9 + 1)
21. (27 + 3) - 9
22. (27 + 1 + 3) - 9
23. 27 - (1 + 3)
24. 27 - 3
25. (27 + 1) - 3
26. 27 - 1
27. 27
28. 27 + 1
29. (27 + 3) - 1
30. 27 + 3
31. 27 + 3 + 1
32. (27 + 9) - (3 + 1)
33. (27 + 9) - 3
34. (27 + 9 + 1) - 3
35. (27 + 9) - 1
36. 27 + 9
37. 27 + 9 + 1
38. (27 + 9 + 3) - 1
39. 27 + 9 + 3
40. 27 + 9 + 3 + 1

**A Basic Baffler**, Page 36
1. L, 2. E, 3. N, 4. E, 5. D, 6. L, 7. W, 8. O
Unscrambled: WELL DONE

**There's No Place Like Rome**, Page 37
Across
1. 3652, 4. 1941, 8. 209, 10. 532, 11. 83, 12. MCC, 14. 60, 15. MMCCX, 17. MCMLXIV, 19. DCVII, 21. 71, 23. VII, 24. 10, 25. 389, 27. 946, 28. 275, 29. 1028
Down
1. 3281, 2. 603, 3. 59, 5. 95, 6. 436, 7. 1207, 9. DCCLVI, 12. MMMCV, 13. CCXII, 15. MCD, 16. XII, 18. 2732, 20. 2068, 22. 187, 24. 142, 26. 95, 27. 90

**Roman Multiplication**, Page 38
1. XX, 2. XXV, 3. CL, 4. XV, 5. XXXVI, 6. CCL, 7. LXXX, 8. XX, 9. XC, 10. CLX

**These Are "Sum" Squares!** Page 39
Variations are possible.

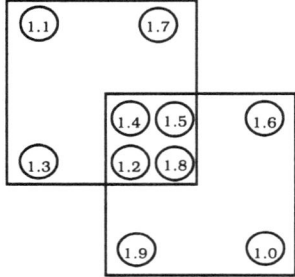

**Start Your Engines!** Pages 40-41
A. 3, B. 10, C. 5, D. 8, E. 1, F. 6, G. 12, H. 7, I. 2, J. 9, K. 4, L. 11
A. Mansell, B. 221.176 mph, C. slower, by .921 mph
Bonus: No, Fittipaldi was the winner.

**House Hassle**, Page 42
1. 589.29 x 1.59 = 936.97
2. 936.97 x 3 = 2810.91
3. 2810.91 + 65.49 + 439.99 + 589.29 = 3905.68
4. 3905.68 x 218.70 = 854,172.22
5. 854,172.22 - 43,999 = 810,173.22
6. $810,173.22 ÷ 8.89 = $91,133.10

**A"maze"ing Mathematics II**, Page 43
Pattern: Answers that increase by 2.0

Start
| 1.1 | 2.1 | 3.1 | 5.1 |
|---|---|---|---|
| 7.1 | 4.1 | 6.1 | 8.1 |
| 9.1 | 14.1 | 12.1 | 10.1 |
| 13.1 | 16.1 | 11.1 | 15.1 |
| 17.1 | 18.1 | 20.1 | 21.1 |

Finish

**Bug Off!** Page 44
They add to misery, subtract from pleasure, divide your attention, and multiply quietly.

**A Quest for Success**, Page 45
Some people may succeed because they are destined to, but most succeed because they are determined to.

**Fractured Cubes**, Page 46
1. $2/3$, 2. $2/7$, 3. $1/4$

**A"maze"ing Mathematics III**, Page 47
Pattern: Answers decrease by $1/4$.
Start

| 3 | $3\frac{1}{4}$ | $3\frac{1}{2}$ | 4 |
|---|---|---|---|
| $2\frac{3}{4}$ | $2\frac{1}{8}$ | $1\frac{3}{4}$ | $1\frac{1}{2}$ |
| $2\frac{1}{2}$ | $2\frac{1}{4}$ | 2 | $1\frac{1}{4}$ |
| $1\frac{1}{2}$ | $1\frac{1}{8}$ | $5/8$ | 1 |
| $9/10$ | $5/6$ | 2 | $3/4$ |

Finish

**Once Is Enough**, Page 48
a. $7/14 + 6/12 = 1$
b. $13/15 + 4/10 = 1\frac{4}{15}$
c. $9/16 - 3/8 = 3/16$
d. $1/2 - 5/11 = 1/22$

**Once Again**, Page 49
a. $2/10 \times 3/8 = 3/40$
b. $13/14 \div 6/12 = 1\frac{6}{7}$
c. $7/16 + 1/4 = 1\frac{3}{4}$
d. $11/15 \times 5/9 = 1\frac{1}{27}$

**"Sum"thing's Strange!** Page 50
a. $7\frac{1}{5}$    d. $4\frac{1}{2}$
   $7\frac{1}{5}$      $4\frac{1}{2}$
b. $4\frac{1}{2}$    e. $3\frac{1}{3}$
   5      $2\frac{1}{3}$
c. $4\frac{1}{3}$    f. $8\frac{1}{6}$
   4

**Other number pairs that work are $1\frac{1}{4} + 5$, $1\frac{1}{7} + 8$, $1\frac{1}{8} + 9$, etc.

**Percentage Predictions**, Page 51
1. G, 2. P, 3. V, 4. B, 5. J, 6. S, 7. D, 8. M, 9. Z, 10. H, 11. T, 12. W, 13. Q, 14. F, Extra Letters: OKAY

**Clancy's Clothing Sale**, Page 52
Explanations may vary.
a. okay
b. The savings is only 17%.
c. If the $49.99 dress were discounted 30%, the sale price should be $34.99.
d. okay
e. Need more information:
If the $21.99 jeans are reduced to $8.80, then there is a savings of 60%. But the ad only states the price is less than $10.
f. The sale prices are only about 15% less.
g. Need more information:
The ad doesn't state the exact sale price of the $18.99 coats. They would have to be $14.24 or less to be reduced by 25%.

**Score More**, Page 54
1. 30% of 90
2. both
3. $5/8$ of 24
4. both
5. $3/4$ of 64
6. both
7. $9/10$ of 90
8. 60% of 300
9. 5% of 90
10. $1/5$ of 120
11. both
12. 10% of 3200

**Sports Court Sale**, Page 55
1. $71.56
2. Soccer ball + hockey stick = $43.05
Soccer ball + football = $32.29
Soccer ball + basketball = $37.80
Football + basketball = $44.89
3. Answers may vary.

**Saver's Savor**, Page 56
a. $28    e. $39
b. $45    f. $6
c. $42    g. $24
d. Jill's plan    h. Joe's plan

**Surprise Graph I**, Page 57

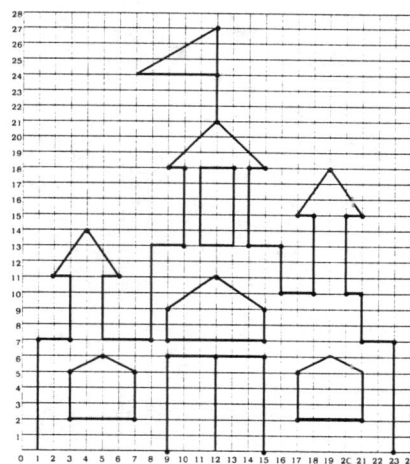

**Surprise Graph II**, Page 59

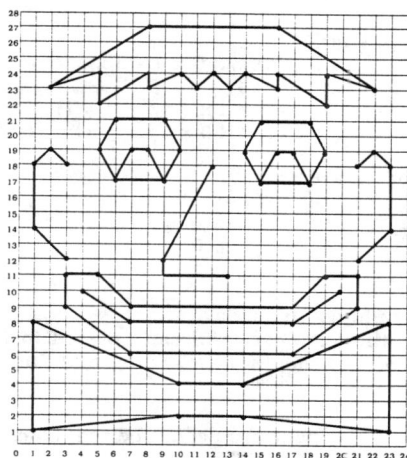

**Map Maker Mo-o-o-ves**, Page 60
a. Udderville and Beefsteak, b. 40 miles

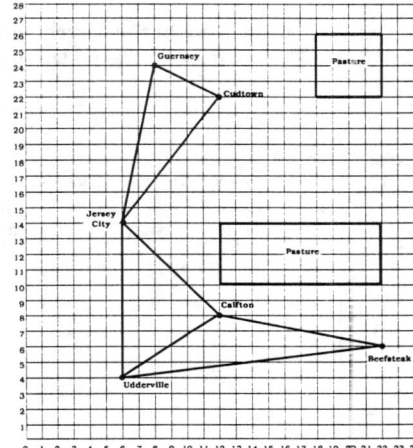

**Putter's Plotting**, Page 61-62
The windmill should be placed by Hole 4.

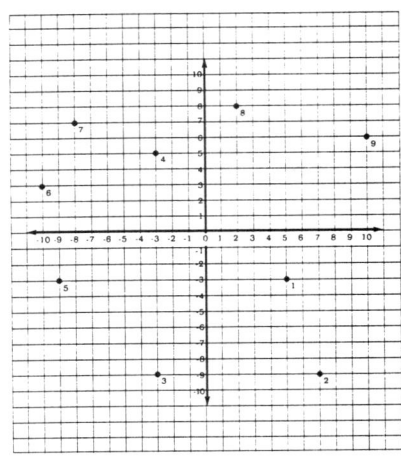

**Sweet Tooth**, Page 63
40 gumdrops, 6 lollipops, 4 candy canes

**Batter Up!** Page 66
1. Shorty McAllister
2. Samuel Strikeout
3. Tommy Triple
4. .270
5. .005
6. .290

**Batter Up!** Page 67
a. 362,880 (9 x 8 x 7 x 6 x 5 x 4 x 3 x 2 x 1)
b. 994.19 years
c. 5,040 (1 x 7 x 6 x 5 x 4 x 3 x 2 x 1 x 1)
d. 144 (1 x 3 x 2 x 1 x 4 x 3 x 2 x 1 x 1)

**April Odds**, Page 68
a. $15/30 = 1/2$
b. $15/30 = 1/2$
c. $4/30 = 2/15$
d. $5/30 = 1/6$
e. $21/30 = 7/10$
f. $12/30 = 2/5$
g. 0

**The New Shoes Blues**, Page 69
a. Since the left shoe is already placed in the correct box, let's assume they are arranged in this manner:

In Box 1, the right shoe could be red, black, green or white. In Box 2 the right shoe could be any of the four remaining colors. In Box 3, the right shoe could be any of the three remaining colors and so on. So the answer is:
5 x 4 x 3 x 2 x 1 = 120.
b. $1/120$
c. Zero! If four right shoes were in the correct boxes, then the fifth one would have to be in its correct box too.

**Square Dare**, Page 70
Outcomes may vary. The key is to make sure each line intersects every other line. The maximum number of sections possible is 11. Here is one way it can be drawn:

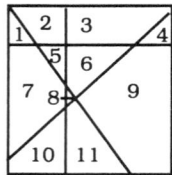

**Box It Up!** Page 71

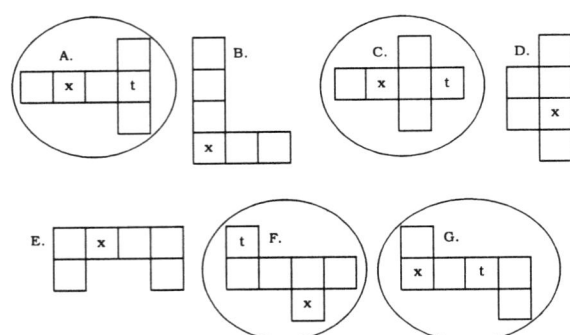

**A Pointed Problem**, Page 72
24 different ways. The arrow can be facing any one of the box's 6 sides. In addition, the arrow can be pointed in any of 4 directions on each side. 6 x 4 = 24

**Trace Case**, Page 73
Shapes A, B, and F

**Network News**, Page 74
Networks A, D, and F

**More Network News**, Page 75
A. No, B. It has four odd vertices.

**Spying on the Spies**, Page 76

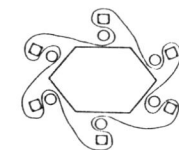

**Square Fare**, Page 77
Cut each shape as shown on the dotted line.

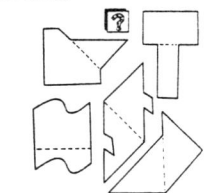

**Cubic Questions**, Page 78
A. 1000, B. 8, C. 96, D. 384, E. 512

**Triangle Trouble**, Page 79
There are eight different triangles shown here. (Numbers 1 and 2 are actually the same shape but of different size.)

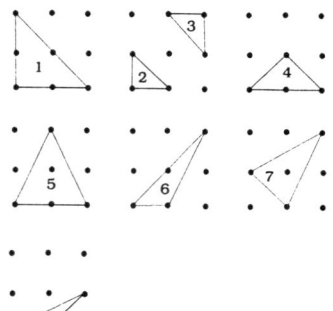

**Area Alert**, Pages 80-81
1. $1/2$ square unit
2. a. 7  b. $4 1/2$  c. 4
3. b. 0, 3, $1/2$  e. 0, 6, 2
   c. 0, 4, 1  f. 0, 12, 5
   d. 0, 10, 4  g. 0, 7, $2 1/2$
4. $A = 1/2 b - 1$ Variations are possible.
5. a. 1, 7, $3 1/2$  d. 1, 10, 5
   b. 1, 8, 4  e. 1, 6, 3
   c. 1, 4, 2  f. 1, 7, $3 1/2$
6. $A = b/2$ Variations are possible.
7. a. 6  d. 12
   b. 8  e. 10
   c. $9 1/2$  f. 11
8. a. 3 sides, a = $1/2$ sq. unit  e. 7 sides, a = $2 1/2$ sq. units
   b. 4 sides, a = 1 sq. unit  f. 8 sides, a = 3 sq. units
   c. 5 sides, a = $1 1/2$ sq. unit  g. 9 sides, a = 4 sq. units
   d. 6 sides, a = 2 sq. units  h. 10 sides, a = $4 1/2$ sq. units

Drawings will vary, but each shape should have 0 interior dots. Here are some examples of solutions.

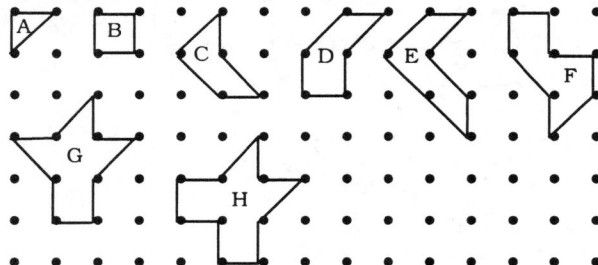

**Fencing Fido and Friends**, Page 82
a. Fido, 288 ft. Sparky, 324 ft. Rover, 294 ft.
b. Fido, 96 yds.; Sparky, 108 yds.; Rover, 98 yds.

**Fencing Finesse**, Page 83
a. 6561 sq. ft.
b. The best solution is to draw a circle.
c. To find the circle's area, we must know its radius. To find that, we divide the circumference (324 ft.) by pi to get the diameter.
   $324 \div 3.1416 = 103.13$ ft.
   Next we divide the diameter by 2 to find the radius.
   $103.13 \div 2 = 51.57$
   Finally we compute the area using the formula $A = \pi r^2$
   $51.57 \times 51.57 \times 3.1416 = 8354.97$ sq. ft.
   That's over 1,790 square feet of additional room for the puppies.
d. approximately 27%

**How Far Is a Furlong?** Page 84
1. C, 2. B, 3. A, 4. C, 5. B, 6. A, 7. C, 8. A, 9. C, 10. B

**Pay Day**, Page 85
a. Students may choose any option, but they will always be paid the most if they select nickels.
b. Answers may vary.
c. approximately: 1 quarter/in., 1.5 dimes/in., 13 nickels stacked/in.
d. Answers may vary. For a student who is 5 ft. tall, the answers would be as follows: $15 in quarters, $9 in dimes, $39 in nickels

**Trivial Lengths and Widths**, Page 86-87
1. C   5. A   9. C   13. B
2. A   6. B   10. A  14. C
3. B   7. C   11. B  15. B
4. A   8. A   12. A  16. A

**Long Gone**, Page 88
1 thru 3. Answers may vary.
4. about 8 or 9 omers
5. about 32
6. 2 ephahs
7. 1 kg

**Billy's Chili**, Page 89
The cylinder, shape B, holds the most, 603 cm³.

**Shapely Numbers I**, Page 90
A.   16            25            36

b. 49, 64, 81, 100, 121
c. Consecutive whole numbers are squared ($x^2$).
e. Rectangles in any of these shapes could be drawn:
   1 x 24, 2 x 12, 3 x 8, 4 x 6
f. There are 6 different shapes:
   1 x 60, 2 x 30, 3 x 20, 4 x 15, 5 x 12, 6 x 10
g. Again, there are 6 different shapes:
   1 x 72, 2 x 36, 3 x 24, 4 x 18, 6 x 12, 8 x 9
h. These would be prime numbers.

**Shapely Numbers II**, Page 91
I.        15

          21

          28

j. 36, 45, 55
k. Add consecutive whole numbers (+2, +3, +4, +5, etc.)

l. 25

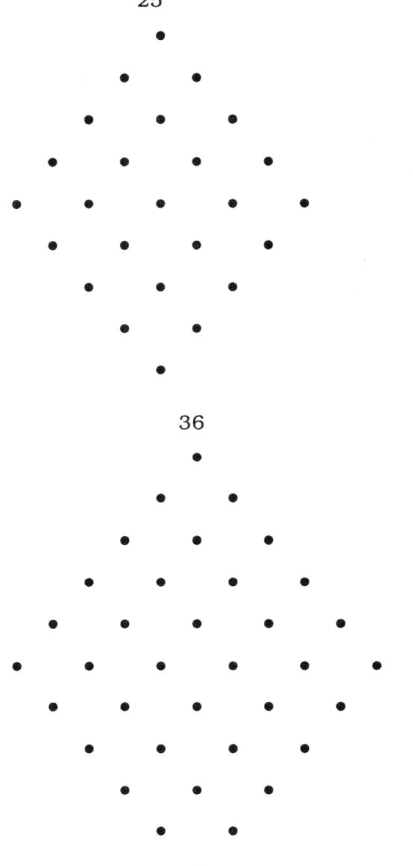

36

m. 49, 64

n. Add consecutive odd numbers (+5, +7, +9, +11, etc.) or square consecutive whole numbers ($2^2$, $3^2$, $4^2$, etc.). The number of dots in the longest (center) row is the number of dots to square to get the total number of dots in the entire diamond. Try to steer students to the observation that a diamond follows the same pattern as the square.

**Campground Confusion**, Page 92
Here is one possible solution. Various answers are possible.

|  | Monday | Tuesday | Wednesday | Thursday | Friday | Saturday |
|---|---|---|---|---|---|---|
| Archery | Cardinals | Orioles | Sparrows | Robins | Eagles | Wrens |
| Canoeing | Wrens | Cardinals | Orioles | Sparrows | Robins | Eagles |
| Horseback Riding | Eagles | Wrens | Cardinals | Orioles | Sparrows | Robins |
| Swimming | Robins | Eagles | Wrens | Cardinals | Orioles | Sparrows |
| Hiking | Sparrows | Robins | Eagles | Wrens | Cardinals | Orioles |
| Softball | Orioles | Sparrows | Robins | Eagles | Wrens | Cardinals |

**Picky, Picky**, Page 93
1. 14 squares
2.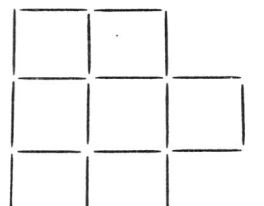
3.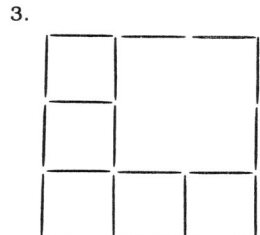

**Tiny Teasers**, Page 94-95
1. Four pieces of paper. Books start with page 1 on the right hand side. Page 2 would be on the back of the same piece of paper. Any pair of consecutive pages starting with an odd number would be on the same page. So pages 35 and 36 are one piece, 54 a second piece, 55 a third, and pages 103 and 104 are on the fourth piece.
2. Four days. Both the men and the amount of work are cut in half, so the rate of work remains the same.
3. Three ducks
4. 114 times, at these 57 times every 12 hours.

| 1:01 | 2:02 | 3:03 | 4:04 | 5:05 | 6:06 | 7:07 | 8:08 | 9:09 |
| 1:11 | 2:12 | 3:13 | 4:14 | 5:15 | 6:16 | 7:17 | 8:18 | 9:19 |
| 1:21 | 2:22 | 3:23 | 4:24 | 5:25 | 6:26 | 7:27 | 8:28 | 9:29 |
| 1:31 | 2:32 | 3:33 | 4:34 | 5:35 | 6:36 | 7:37 | 8:38 | 9:39 |
| 1:41 | 2:42 | 3:43 | 4:44 | 5:45 | 6:46 | 7:47 | 8:48 | 9:49 |
| 1:51 | 2:52 | 3:53 | 4:54 | 5:55 | 6:56 | 7:57 | 8:58 | 9:59 |

10:01 11:11 12:21
5. 41,312,432
6.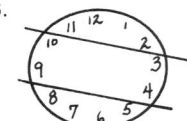

7. If your line is long enough, it will take only fifty minutes. They can dry simultaneously.
8. Shari - 8, 10, 12 = 30 oz.
   Mary - 4, 5, 6, = 15 oz.
   Teri - 9, 13, 15 = 37 oz.

**Publishers' Paradise**, Page 96

| 9:00 - 10:00 | Meet with editor |
| 10:00 - 12:00 | Meet with marketing director |
| 12:00 | Meet with secretary |
| 1:00 | Meet with book designer |
| 2:00 | Break |
| 3:00 - 4:00 | Meet with illustrator |

**Puzzling Pizzeria**, Page 97
I ordered a 14" cheese and pepperoni pizza and a 16" cheese, pepperoni, and mushroom pizza.

**Zams for Sale I**, Page 98
The xob is 3 geef x 3 geef x 2 geef, which equals 18 cubic geef. Since each Zam is 1 cubic geef, the xob will hold 18 Zams.

**Zams for Sale II**, Page 99
5 fruit cakes are worth 15 zucchini, which added to Sophie's 5 totals 20 zucchini. Since every 4 zucchini can buy one Zam, Sophie can buy 5 Zams.

**Rope Riddle**, Page 100
Use your rope to find a length of 7 yards by measuring the length of the shed. Then measure the width of the shed to find a length of rope 5 yards. Compare your two rope lengths. The difference is 2 yards. Now compare your 2-yard length with the 5-yard length. The difference is a 3-yard length. Finally, add your 5-yard length to your 3-yard length and you will have an 8-yard length of rope with which to measure your mother's garden.

**Christmas Candy Confusion**, Page 101
Number your boxes from 1 to 6. Take 1 candy from Box 1, 2 candies from Box 2, 3 from Box 3, etc. Since you have 21 pieces of candy now removed from the 6 boxes, the weight shown on the scales should read 210 grams. The weight shown on the scale when you weigh all 21 pieces will show you which box is lighter. For example, if the scale is 1 gram short (or 209g), you know that only 1 candy is lighter which had to have come from Box 1. If the scales show 205 g, which is 5 grams lighter than expected, then you have 5 pieces which are smaller. Those came from Box 5, and so on.